Preparation and Application of Polymer Nanocomposites

Preparation and Application of Polymer Nanocomposites

Editor

Teresa Cuberes

MDPI • Basel • Beijing • Wuhan • Barcelona • Belgrade • Manchester • Tokyo • Cluj • Tianjin

Editor
Teresa Cuberes
University of Castilla-La Mancha
Spain

Editorial Office
MDPI
St. Alban-Anlage 66
4052 Basel, Switzerland

This is a reprint of articles from the Special Issue published online in the open access journal *Nanomaterials* (ISSN 2079-4991) (available at: https://www.mdpi.com/journal/nanomaterials/special_issues/Polymer_Nanocomposites_Application).

For citation purposes, cite each article independently as indicated on the article page online and as indicated below:

LastName, A.A.; LastName, B.B.; LastName, C.C. Article Title. *Journal Name* **Year**, *Volume Number*, Page Range.

ISBN 978-3-0365-7122-5 (Hbk)
ISBN 978-3-0365-7123-2 (PDF)

Cover image courtesy of Prof. Dr. Teresa Cuberes

© 2023 by the authors. Articles in this book are Open Access and distributed under the Creative Commons Attribution (CC BY) license, which allows users to download, copy and build upon published articles, as long as the author and publisher are properly credited, which ensures maximum dissemination and a wider impact of our publications.
The book as a whole is distributed by MDPI under the terms and conditions of the Creative Commons license CC BY-NC-ND.

Contents

Teresa Cuberes
Preparation and Application of Polymer Nanocomposites
Reprinted from: *Nanomaterials* **2023**, *13*, 657, doi:10.3390/nano13040657 1

Zhibin Ren, Yongqiang Zhu, Qi Wu, Minye Zhu, Feng Guo, Huayang Yu and Jiangmiao Yu
Enhanced Storage Stability of Different Polymer Modified Asphalt Binders through Nano-Montmorillonite Modification
Reprinted from: *Nanomaterials* **2020**, *10*, 641, doi:10.3390/nano10040641 3

Sakhayana N. Danilova, Afanasy A. Dyakonov, Andrey P. Vasilev, Aitalina A. Okhlopkova, Aleksei G. Tuisov, Anatoly K. Kychkin and Aisen A. Kychkin
Effect of Borpolymer on Mechanical and Structural Parameters of Ultra-High Molecular Weight Polyethylene
Reprinted from: *Nanomaterials* **2021**, *11*, 3398, doi:10.3390/nano11123398 21

Georgia Geka, George Papageorgiou, Margarita Chatzichristidi, Andreas Germanos Karydas, Vassilis Psycharis and Eleni Makarona
CuO/PMMA Polymer Nanocomposites as Novel Resist Materials for E-Beam Lithography
Reprinted from: *Nanomaterials* **2021**, *11*, 762, doi:10.3390/nano11030762 39

Liang Zhao and Chun Liang Liu
Review and Mechanism of the Thickness Effect of Solid Dielectrics
Reprinted from: *Nanomaterials* **2020**, *10*, 2473, doi:10.3390/nano10122473 61

Gen-Wen Hsieh, Liang-Cheng Shih and Pei-Yuan Chen
Porous Polydimethylsiloxane Elastomer Hybrid with Zinc Oxide Nanowire for Wearable, Wide-Range, and Low Detection Limit Capacitive Pressure Sensor
Reprinted from: *Nanomaterials* **2022**, *12*, 256, doi:10.3390/nano12020256 85

Pavel Tofel, Klára Částková, David Říha, Dinara Sobola, Nikola Papež, Jaroslav Kaštyl, Ştefan Ţălu, and Zdeněk Hadaš
Triboelectric Response of Electrospun Stratified PVDF and PA Structures
Reprinted from: *Nanomaterials* **2022**, *12*, 349, doi:10.3390/nano12030349 99

Aleksandra Wesełucha-Birczyńska, Anna Kołodziej, Małgorzata Świetek, Łukasz Skalniak, Elżbieta Długoń, Maria Pajda and Marta Błażewicz
Early Recognition of the PCL/Fibrous Carbon Nanocomposites Interaction with Osteoblast-like Cells by Raman Spectroscopy
Reprinted from: *Nanomaterials* **2021**, *11*, 2890, doi:10.3390/nano11112890 113

Joao Augusto Oshiro Jr., Angelo Lusuardi, Elena M. Beamud, Leila Aparecida Chiavacci and M. Teresa Cuberes
Nanostructural Arrangements and Surface Morphology on Ureasil-Polyether Films Loaded with Dexamethasone Acetate
Reprinted from: *Nanomaterials* **2021**, *11*, 1362, doi:10.3390/nano11061362 131

Editorial
Preparation and Application of Polymer Nanocomposites

Teresa Cuberes

Department of Applied Mechanics and Project Engineering, Mining and Industrial Engineering School of Almaden, University of Castilla-La Mancha, Plaza Manuel Meca 1, 13400 Almadén, Spain; teresa.cuberes@uclm.es

The incorporation of nanomaterials into polymer matrices opens new avenues for the development of advanced materials with unique novel properties and impact in many different fields. A thorough understanding of how nanofillers affect the structural conformation of the polymer matrix at different hierarchical levels; how to control the dispersion, aggregation, and organization of nanofillers within the matrix; and how to tune and monitor the interfacial properties still remains a challenge. The current issue includes interesting examples of the preparation of polymer nanocomposites for different applications:

Structural nanocomposite materials and their mechanical properties are considered in [1,2]. In [1], the benefits of the incorporation of nanoclay into polymer-modified asphalt are investigated; intercalated or exfoliated nanoclay structures hinder the migration of insoluble additive of polymers, hindering phase separation, and improving its storage stability, which is a major concern in asphalt modification technologies. [2] studies the effect of adding borpolymer to ultra-high-molecular-weight polyethylene, which does not chemically interact with the matrix but acts as a reinforcing filler, modifying the supramolecular structure of the matrix, and hence its deformation and strength responses.

Nanocomposites for electronics and nanofabrication are considered in [3,4]. In [3], an innovative resist based on CuO/polymethyl methacrylate (PMMA) is developed; the CuO nanostructures with typical sizes of 10–30 nm provide functionality to the resist, which can be patterned by electron beam lithography. [4] reviews the dielectric-thickness dependence of the electric breakdown strength, which is a critical parameter in the design of solid insulation structures, discussing the responsible mechanisms for the thickness effect.

Nanocomposite-based nanosensors and nanogenerators are considered in [5,6]. In [5], a flexible capacitive pressure sensor is proposed, formed by a nanocomposite dielectric layer of porous polydimethyl siloxane elastomer and ZnO nanowire, which maintains its response over a wide dynamic range from 1 Pa to 50 KPa, almost covering the entire tactile-pressure range. Ref. [6] develops triboelectric energy harvesters based on polyvinylidene fluoride (PVDF) and polyvinyl amide (PA) fibrous composites, evaluating their dielectric and triboelectric responses.

Eventually, nanocomposites for biomedical applications are considered in [7,8]. In [7], poly(ε-caprolactone) nanocomposites with multiwall nanotubes and electrospan carbon fibers are investigated to improve cell adhesion; human osteoblast-like cells were successfully developed on the nanocomposites, with their interactions being conveniently monitored from the early stages through Raman microspectroscopy. Ref. [8] studies organic–inorganic ureasil–polyether hybrid matrices for drug delivery, using dexamethasone acetate as a model drug; ultrasonic force microscopy reveals a structural organization at ureasil–poly(propylene oxide)400 films in which clusters of correlated siloxane nodes form beads that align into strands, gathering into hybrid polymer ropes on the film surface, which impacts the behavior of the matrix as a drug host.

Conflicts of Interest: The author declares no conflict of interest.

References

1. Ren, Z.; Zhu, Y.; Wu, Q.; Zhu, M.; Guo, F.; Yu, H.; Yu, J. Enhanced Storage Stability of Different Polymer Modified Asphalt Binders through Nano-Montmorillonite Modification. *Nanomaterials* **2020**, *10*, 641. [CrossRef] [PubMed]
2. Danilova, S.N.; Dyakonov, A.A.; Vasilev, A.P.; Okhlopkova, A.A.; Tuisov, A.G.; Kychkin, A.K.; Kychkin, A.A. Effect of Borpolymer on Mechanical and Structural Parameters of Ultra-High Molecular Weight Polyethylene. *Nanomaterials* **2021**, *11*, 3398. [CrossRef]
3. Geka, G.; Papageorgiou, G.; Chatzichristidi, M.; Karydas, A.G.; Psycharis, V.; Makarona, E. CuO/PMMA Polymer Nanocomposites as Novel Resist Materials for E-Beam Lithography. *Nanomaterials* **2021**, *11*, 762. [CrossRef] [PubMed]
4. Zhao, L.; Liu, C.L. Review and Mechanism of the Thickness Effect of Solid Dielectrics. *Nanomaterials* **2020**, *10*, 3473. [CrossRef] [PubMed]
5. Hsieh, G.-W.; Shih, L.-C.; Chen, P.-Y. Porous Polydimethylsiloxane Elastomer Hybrid with Zinc Oxide Nanowire for Wearable, Wide-Range, and Low Detection Limit Capacitive Pressure Sensor. *Nanomaterials* **2022**, *12*, 256. [CrossRef] [PubMed]
6. Tofel, P.; Částková, K.; Říha, D.; Sobola, D.; Papež, N.; Kaštyl, J.; Ţălu, Ş.; Hadaš, Z. Triboelectric Response of Electrospun Stratified PVDF and PA Structures. *Nanomaterials* **2022**, *12*, 349. [CrossRef] [PubMed]
7. Wesełucha-Birczyńska, A.; Kołodziej, A.; Świętek, M.; Skalniak, Ł.; Długoń, E.; Pajda, M.; Błażewicz, M. Early Recognition of the PCL/Fibrous Carbon Nanocomposites Interaction with Osteoblast-like Cells by Raman Spectroscopy. *Nanomaterials* **2021**, *11*, 2890. [CrossRef] [PubMed]
8. Oshiro-Junior, J.A.; Lusuardi, A.; Beamud, E.M.; Chiavacci, L.A.; Cuberes, M.T. Nanostructural Arrangements and Surface Morphology on Ureasil-Polyether Films Loaded with Dexamethasone Acetate. *Nanomaterials* **2021**, *11*, 1362. [CrossRef] [PubMed]

Disclaimer/Publisher's Note: The statements, opinions and data contained in all publications are solely those of the individual author(s) and contributor(s) and not of MDPI and/or the editor(s). MDPI and/or the editor(s) disclaim responsibility for any injury to people or property resulting from any ideas, methods, instructions or products referred to in the content.

Article

Enhanced Storage Stability of Different Polymer Modified Asphalt Binders through Nano-Montmorillonite Modification

Zhibin Ren [1,2,3], Yongqiang Zhu [4], Qi Wu [5], Minye Zhu [2], Feng Guo [6], Huayang Yu [3,*] and Jiangmiao Yu [2,*]

1. Research and Development Centre of Transport Industry of Technologies, Materials and Equipment of Highway Construction and Maintenance, Gansu Road and Bridge Group Co. Ltd., Lanzhou 730050, China; mszhibinren@mail.scut.edu.cn
2. School of Civil Engineering and Transportation, South China University of Technology, Wushan Road, Tianhe District, Guangzhou 510000, China; 201730031412@mail.scut.edu.cn
3. Key Laboratory for Special Area Highway Engineering, Chang'an University, Xi'an 710064, China
4. Guangdong Guanyue Highway & Bridge Co., Ltd., Guangzhou 510000, China; ZHU15602229920@126.com
5. Guangdong Province Communications Planning & Design Institute Co., Ltd., Xinghua Road, Tianhe District, Guangzhou 510000, China; wuqi614@126.com
6. Department of Civil Engineering and Environment, University of South Carolina, Columbia, SC 29208, USA; fengg@email.sc.edu
* Correspondence: huayangyu@scut.edu.cn (H.Y.); yujm@scut.edu.cn (J.Y.); Tel.: +86-020-87111030 (H.Y.)

Received: 6 March 2020; Accepted: 21 March 2020; Published: 30 March 2020

Abstract: The storage stability concern, caused by phase separation for the density difference between polymers and asphalt fractions, has limited the widespread application of polymer modified asphalt (PMA). Therefore, this study aims to improve the storage concern of PMA by incorporating nano-montmorillonite. To this end, different nano-montmorillonites were incorporated to three PMAs modified with three typical asphalt modifiers, i.e., crumb rubber (CRM), styrene–butadiene-rubber (SBR) and styrene–butadiene-styrene (SBS). A series of laboratory tests were performed to evaluate the storage stability and rheological properties of PMA binders with nano-montmorillonite. As a consequence, the incorporation of nano-montmorillonite exhibited a remarkable effect on enhancing the storage stability of the CRM modified binder, but limited positive effects for the SBR and SBS modified binders. The layered nano-montmorillonite transformed to intercalated or exfoliated structures after interaction with asphalt fractions, providing superior storage stability. Among selected nano-montmorillonites, the pure montmorillonite with Hydroxyl organic ammonium performed the best on enhancing storage stability of PMA. This paper suggests that nano-montmorillonite is a promising modifier to alleviate the storage stability concern for asphalt with polymer modifiers.

Keywords: storage stability; rheological properties; polymer-modified asphalt; nano-montmorillonite

1. Introduction

With the extremely increasing loading and aggravation of axis load in the pavement industry, damages such as rutting, cracking etc., are happening more frequently on highways and urban roads [1–4]. Asphalt modification technology has been considered a practical approach to resolve these concerns by enhancing the durability of asphalt pavements. Crumb rubber (CRM), styrene-butadiene-rubber (SBR) and styrene-butadiene-styrene (SBS) are the three most widely applied modifiers, which are regarded as effective adhesive and cohesive performance enhancers of asphalt [5–9]. However, the storage stability concern of modified asphalt has limited its widespread application.

Engineers from asphalt plants have always worried about the separation of the modifier and asphalt during the storage and transportation process in elevated temperatures.

According to Stoke's law, the phase separation phenomenon can be governed by the following equation [10–12]:

$$v_t = \frac{2a^2 \Delta \rho g}{9\eta} \tag{1}$$

where v_t is the settling velocity of dispersed particles, a is the radius of dispersed particles, $\Delta \rho$ is the density difference between two different phases, g is gravitational acceleration, and η is the dynamic viscosity of liquid medium.

Among different asphalt modifiers, CRM tends to sink in the liquid phase of modified asphalt during the storage process due to higher density compared to virgin asphalt, while SBR and SBS additives with lower density values tend to float in the upper part of liquid phase. The separation of modifier and raw asphalt results in a huge difference in composition and rheological properties between top and bottom portions of the polymer modified asphalt after storage. Previous studies indicate that this concern can be alleviated by adjusting the liquid asphalt density using bio-modification [13,14] or activating the crumb rubber [15]. However, the improvement effect is not very satisfactory. Therefore, one potential method is addressed in this study by incorporating nanoclay into the polymer-modified asphalt, which can reduce the phase separation phenomenon by decreasing the migration velocity of insoluble additive of polymers in the liquid phase [16,17].

Nanoclay is a type of natural mineral mainly including kaolinite clay (KC), vermiculite (VMT) and montmorillonite (MMT), which has a 2:1 layered structure with two silica tetrahedral sheets sandwiching an alumina octahedral sheet. Nowadays, nanomaterials have been popularly applied as modifiers for construction materials [12,18–20]. Especially, the importance of layered clay minerals, also known as nanoclays, in terms of asphalt modification is gradually increasing. Nanoclays' remarkable improvement on the rheological properties of asphalt has been widely reported by previous studies. Vargas et al., [18] discovered that Organo-nanocomposite modified-asphalt can generate an intercalated structure using X-ray diffraction (XRD) and transmission electron microscopy (TEM). This indicates that the enhanced rheological properties may ascribe to the interaction behavior of the polymer chains in asphalt binder into the interlayer of clay. Yu et al., [21] enhanced the storage stability of asphalt rubber by incorporating three types of nanoclays and improved the rheological properties through modification. Leng et al., [22] found that clay/SBS modified bitumen composites have acceptable storage performance. The composites also showed better resistance to aging by reducing the oxidation of bitumen and the degradation of SBS. Galooyak et al., [23] proved the improvement of nanoclays on the storage stability of SBS modified asphalt and confirmed the conclusion through morphological analysis. Thus, nanoclay was expected by the asphalt industry to serve as the storage stability improver of PMAs.

The objective of this study was to evaluate the feasibility of alleviating the storage stability concern of polymer-modified asphalt by incorporating nano-montmorillonite. To this end, three different types of nano-montmorillonites and three types of modifiers, i.e., CRM, SBR and SBS, were selected to prepare NPMAs. The rheological tests were performed to evaluate mechanical properties of NPMA binders when applied to pavement industry, among which the Superpave rutting factor test was chosen to control the content of these three modifiers. With the same PG82 grade, the content of modifier applied in each PMA binder was high enough to simulate the most unfavorable situation after storing in high temperature. In addition, the storage stability of modified binders was quantitatively analyzed through characterizing the differences in softening point, complex moduli and absorbance ratio of CRM/SBR/SBS in the infrared spectrum between the top and bottom portions of the sample after a lab-simulated storage process. Finally, an X-ray diffraction (XRD) test was conducted to investigate the layer gap distance variation of nano-montmorillonite to reveal the modification mechanism.

2. Materials and Methods

2.1. Materials and Sample Preparation

A Pen 60/70 virgin asphalt, with a penetration grade of 60/70, was used to prepare NPMA binders. In this study, all modified binders including polymer-modified asphalts, nano-montmorillonite-modified asphalts (NMA) and nano-montmorillonite-polymer modified asphalts were prepared by 10,000 rpm high shear incorporating modifiers with a certain dosage into virgin asphalt at 180 °C for 1 h. The selected dosages were 20 wt %, 7 wt % and 6 wt % by virgin asphalt for CRM (40-mesh), SBR and SBS, respectively, then a certain dosage of nano-montmorillonite (3 wt % by virgin asphalt) was adopted to prepare NPMA binders.

Three different types of nano-montmorillonites, labelled as A, B and C, were applied for asphalt modification. The nano-montmorillonite samples used in this study are organomodified nanoclay particles (provided by the Zhejiang Fenghong Clay Chemical Co., Ltd., Huzhou, China). The nanoclay samples are high-quality montmorillonite with high purity (at least 95% montmorillonite content). Besides, the nano-montmorillonite layer was forgeable due to the large specific surface area (750 m^2/g) and the unique layered one-dimensional nanostructure and morphology. Among the three nanoclays, A is pure montmorillonite with Na$^+$ inorganic group, while B and C are montmorillonites having inorganic groups exchanged with different alkyl ammonium ions. The ranking of their surface hydrophilic properties from high to low is A, B and C. Different from other two-dimensional and three-dimensional inorganic nanoparticles, the specific structure and morphology might lead to excellent mechanical properties, thermal properties, functional properties and physical properties of nano-montmorillonite-polymer modified asphalts. The morphologies of nano-montmorillonites were presented in Figure 1, and the physical properties of different nano-montmorillonites were shown in Table 1. Table 2 detailed the information of each test sample.

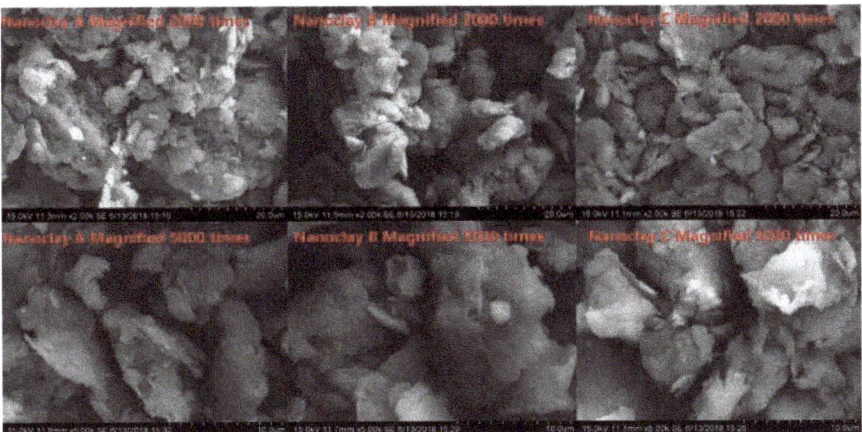

Figure 1. Nano-montmorillonite morphology.

Table 1. Properties of nano-montmorillonite additives.

ID	Nanoclay A	Nanoclay B	Nanoclay C
Modified method	Pure MT with Na$^+$ inorganic group	Hydroxyl organic ammonium	Double alkyl ammonium
Montmorillonite content		96~98%	
Specific gravity	1.8	1.8	1.7
Bulk gravity	<0.3	≤0.3	≤0.3
Hydrophilic	Medium	Strong	Poor
X-ray d001	2.24 nm	2.09 nm	3.73 nm
Applicable polymer	PE, PP, PVC	N/A	PP and other thermos plasticity polymers

Table 2. Composition of different modified asphalt samples.

Binder Type	Sample ID	Type of Modifier	Dosage of Modifier	Type of Nano-Montmorillonite
Virgin asphalt	Pen60/70	N/A	N/A	N/A
NMA	VB-A	N/A	N/A	A
NMA	VB-B	N/A	N/A	B
NMA	VB-C	N/A	N/A	C
PMA	CRM-0	Crumb rubber	20 wt %	N/A
PMA	SBR-0	Styrene-butadiene-rubber	7 wt %	N/A
PMA	SBS-0	Styrene–butadiene-styrene	6 wt %	N/A
NPMA-A	CRM-A	Crumb rubber	20 wt %	A
NPMA-A	SBR-A	Styrene-butadiene-rubber	7 wt %	A
NPMA-A	SBS-A	Styrene–butadiene-styrene	6 wt %	A
NPMA-B	CRM-B	Crumb rubber	20 wt %	B
NPMA-B	SBR-B	Styrene-butadiene-rubber	7 wt %	B
NPMA-B	SBS-B	Styrene–butadiene-styrene	6 wt %	B
NPMA-C	CRM-C	Crumb rubber	20 wt %	C
NPMA-C	SBR-C	Styrene-butadiene-rubber	7 wt %	C
NPMA-C	SBS-C	Styrene–butadiene-styrene	6 wt %	C

2.2. Testing Program

The conventional physical properties, including penetration and softening point, were selected as the indicators for the general properties of test binders. The workability was evaluated using a Brookfield viscometer (RVD VII+) through measuring rotational viscosities of asphalt specimen at three different temperatures.

The rheological properties of modified binders were characterized using a dynamic shear rheometer (DSR, Malvern Kinexus Lab+, Malvern analytical Company, UK). The Superpave rutting factor ($G^*/\sin \delta$) test and multiple stress creep recovery (MSCR) tests were conducted to evaluate the high temperature rutting resistance of asphalt samples, while the intermediate temperature fatigue resistance was analyzed through the Superpave fatigue factor ($G^*\sin \delta$) test and linear amplitude sweep (LAS) test. The test binders for the MSCR test were aged by the standard rolling thin film oven (RTFO) process, while those for the fatigue test were aged by both RTFO and pressure aging vessel (PAV) processes. The bending beam rheometer (BBR) test was also performed to evaluate the low temperature cracking resistance performance of the RTFO + PAV aged samples.

The storage stability of modified binders was quantitatively analyzed through characterizing the differences in softening point [16,17], complex moduli [24–26] and absorbance ratio of CRM/SBR/SBS in the infrared spectrum between the top and bottom portions of the sample after storing. For FTIR tests, the test binder with a thickness of approximately 1 mm was placed in a transmission holder and scanned in order to obtain infrared spectroscopy ranging from 4,000 to 400 cm^{-1}. According to ASTM 7173 [27], to simulate the high temperature storing process in the laboratory, about 70 g of hot asphalt was poured into an aluminum tube with a diameter of 25 mm. Before cutting the tube into three equal parts horizontally, it was being stored at 163 °C for 48 h followed by cooling down at −5 °C.

To investigate the modification mechanism of nano-montmorillonite on storage stability, X-ray diffraction (XRD) tests were conducted to investigate the layer gap distance variation. Table 3 shows the detailed information of conducted tests in this study.

Table 3. Details of the laboratory test.

Performance	Tests	Aging Level	Specification/Standard	Temperature
General properties	Penetration	unaged	ASTM D5	25 °C
	Softening point		ASTM D36	N/A
Workability	Rotational viscosity	unaged	AASHTO T316	135 °C, 150 °C and 165 °C
Rutting resistance	Rutting factor (G*/sin δ)	unaged	AASHTO M320	64–88 °C
	MSCR	RTFO aged	AASHTO MP19-10	64 °C
Fatigue resistance	Fatigue factor (G*sin δ)	RTFO + PAV aged	AASHTO M320	25–13 °C
	LAS		AASHTO TP101	25 °C
Storage stability	Softening point	unaged	ASTM D36	N/A
	Complex shear modulus		AASHTO M320	25 °C, 64 °C and 82 °C
	FTIR		N/A	25 °C
Low temperature cracking resistance	BBR	RTFO + PAV aged	AASHTO T313	−6 °C, −12 °C and −18 °C
Internal layer distance of nano-montmorillonite	XRD	unaged	N/A	25 °C

3. Results and Discussion

3.1. Physical Properties

Figure 2 presents the penetration and softening point results of test binders. It is noted that the polymer-modified asphalts had a higher softening point and a lower penetration, which indicates that the incorporation of CRM/SBR/SBS led to superior performance at high temperature and higher stiffness respectively. It is also observed that the incorporation of nano-montmorillonite further decreased the penetration of SBS-0, while had insignificant effect on CRM-0 and SBR-0. It indicates that the stiffness of SBS modified asphalt binder is more sensitive to nano-montmorillonite compared to the other polymer modified asphalt binders. Different from the penetration results, adding nano-montmorillonite increased the softening points of all modified asphalts.

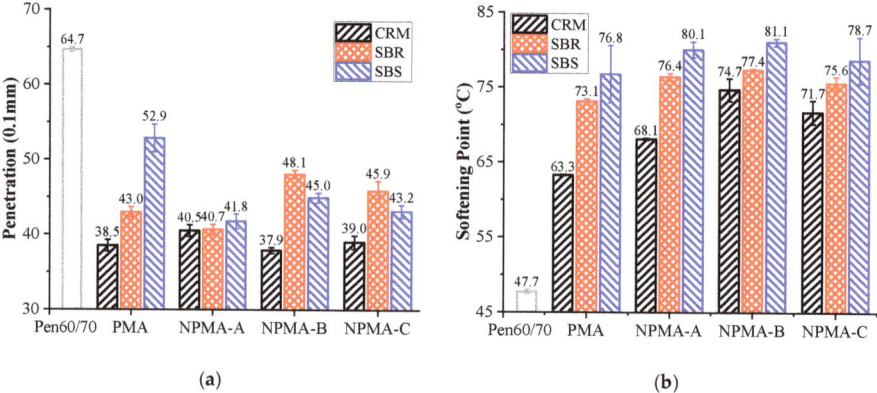

Figure 2. Physical properties of asphalt binders: (**a**) Penetration; (**b**) Softening point.

3.2. Workability

Figure 3 presents the workability results, which were evaluated by the Brookfield rotational viscosity tests. The higher the viscosity value was, the worse workability the test binder had. According to previous studies [28–30], poor workability is one of the most critical concerns limiting the spread of asphalt rubber. As expected, CRM modified asphalt binders had extremely higher viscosities than other test binders at all temperatures. Besides, the viscosities of SBR/SBS modified binders except SBS-C were below 3000 cP at 135 °C, which indicates the mixtures with these binders can be compacted according to the AASHTO (American Association of State Highway and Transportation Officials) specification.

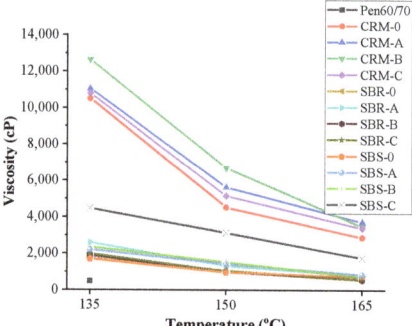

Figure 3. Rotational viscosity test results.

3.3. Rutting Performance

Figure 4b shows the rutting factor (G*/sin δ) values at different temperatures. The starting temperature was set as 64 °C, then the test temperature was automatically increased by 6 °C until the rutting factor was below 1.0 kPa (the critical value for unaged binders). Figure 4a presents the critical temperature results. The higher the critical temperature was, the superior rutting resistance the test binder had. As expected, all of the three polymer modifiers led to much higher critical temperatures than virgin asphalt. It is noted that adding nano-montmorillonite had insignificant effect on SBR-0 and SBS-0. However, the incorporation of nano-montmorillonite further enhanced the rutting resistance of CRM-0, indicating that nano-montmorillonite worked much better with CRM modified binders than SBR/SBS modified binders in terms of rutting resistance.

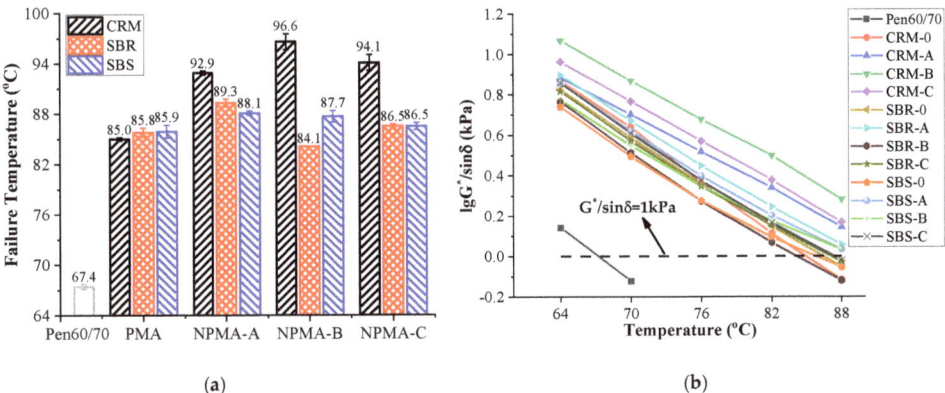

Figure 4. Superpave rutting factor test results: (**a**) Failure temperatures; (**b**) Logarithm of G*/sin δ values.

The MSCR test results were presented in Table 4. Lower J_{nr} values and higher % Recovery values refers to superior pavement rutting resistance performance. It is noted that SBS modified binders with nano-montmorillonite and all CRM modified binders did not meet the requirement of AASHTO specification, i.e., <75%. This can be attributed to the extremely low J_{nr} value at 0.1 kPa. Consistent with the G*/sin δ results, the CRM modified binders exhibited the best rutting property for its lower J_{nr}3.2 value compared to asphalt binders modified with SBS or SBR.

Table 4. MSCR test results.

Binder Type	J_{nr}		J_{nr}% Diff	% Recovery	
	0.1 kPa (kPa^{-1})	3.2 kPa (kPa^{-1})		0.1 kPa (kPa^{-1})	3.2 kPa (kPa^{-1})
Pen60/70	3.988	4.586	15.0	2.1	−0.5
CRM-0	0.019	0.098	402.6	91.6	63.0
CRM-A	0.034	0.168	392.7	89.0	55.8
CRM-B	0.013	0.090	608.0	94.6	67.1
CRM-C	0.023	0.141	518.4	92.0	59.5
SBR-0	0.360	0.417	15.9	34.0	40.5
SBR-A	0.202	0.310	53.7	67.6	54.9
SBR-B	0.310	0.447	44.3	59.6	50.4
SBR-C	0.320	0.463	44.5	52.1	41.8
SBS-0	0.217	0.382	76.2	68.5	51.7
SBS-A	0.060	0.269	350.1	90.5	64.1
SBS-B	0.096	0.300	213.0	88.0	68.1
SBS-C	0.093	0.300	221.9	86.9	63.5

3.4. Fatigue Performance

The fatigue critical temperature results are presented in Figure 5a, while the variation of fatigue factor (G*sin δ) with temperatures are shown in Figure 5b. Similar to the G*/sin δ test process, the starting temperature was set as 25 °C, then the test temperature was automatically decreased by 3 °C until the fatigue factor exceeded 5,000 kPa. According to the AASHTO specification, lower critical temperature indicates better fatigue resistance performance. It is noted that incorporating CRM effectively decreased the failure temperatures of neat asphalt by 6.3 °C. What's more, the critical temperatures of CRM-A and CRM-B were 1.2 and 2.2 lower than that of CRM-0, which indicates that the application of nanoclays A and B can further enhance the fatigue performance of CRM-0. However, the incorporation of SBR and SBS had insignificant and even limited negative effects on the fatigue performance of base binder.

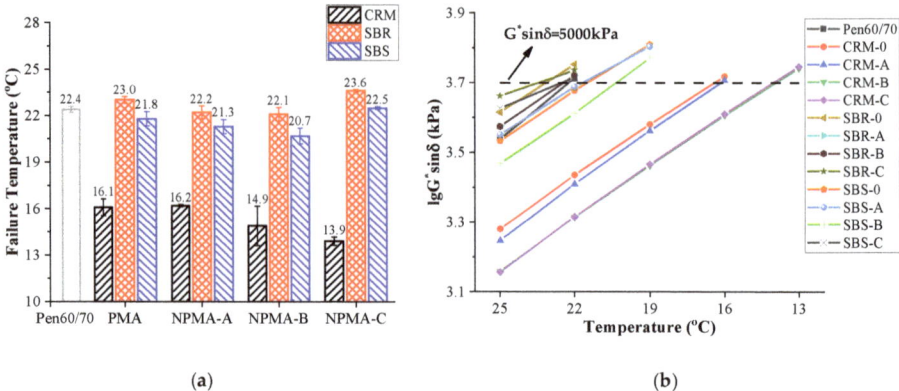

Figure 5. Superpave fatigue factor test results: (**a**) Failure temperatures; (**b**) Logarithm of G*sin δ values.

Figure 6a,b present the fatigue lives (N_f) of LAS tests at 2.5% and 5.0% applied strain levels, respectively. The higher the fatigue life was, the better fatigue cracking resistance the test specimen had. As shown in Figure 6a, it is noted that the N_f of CRM-0, SBR-0 and SBS-0 was 14.5, 2.3 and 3.5 times that of neat asphalt, respectively. This indicates a different test result from the G*sin δ test results that all modified binders performed better than neat asphalt. According to previous studies [31,32], LAS was proven to be a more reliable method for characterizing fatigue performance of asphalt binders. Therefore, it is believed that CRM exhibited outstanding performance in enhancing fatigue resistance performance, while SBR and SBS had limited positive effect on the fatigue performance of neat asphalt. However, the conclusion of fatigue performance is recommended to be validated by mixture tests like the indirect tensile fatigue (ITFT) test and four-point bending beam (4PB) test.

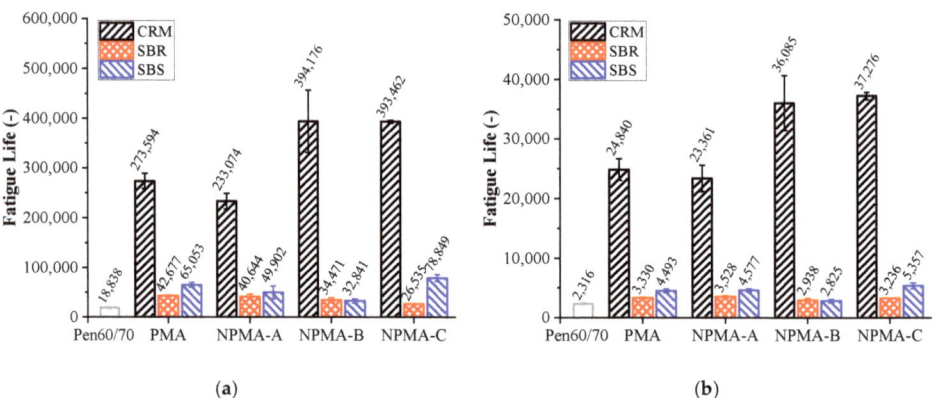

Figure 6. LAS test results: (**a**) Applied strain of 2.5%; (**b**) Applied strain of 5.0%.

3.5. Low Temperature Cracking Performance

Table 5 presents the stiffness and m-values of test binders determined by the BBR test. To meet a specific requirement of AASHTO T313, the stiffness value should be less than 300 MPa, and the m-value should be higher than 0.3. A low-temperature cracking was more likely caused by higher stiffness values. As expected, the incorporation of CRM/SBR/SBS succeeded in enhancing the cracking resistance of neat asphalt by decreasing the stiffness value. Among three PMA binders, the best low temperature cracking resistance was obtained by CRM-0 for the lowest stiffness. It can also be seen that the incorporation of nano-montmorillonites further decreased the stiffness values of all PMA binders. Eventually, all modified binders were compliant with the requirement at −12 °C. Therefore, all three kinds of modifiers and three nano-montmorillonites can contribute to the application of asphalt pavement in colder regions. It is worth mentioning that recent studies show that the cooling period, temperature and medium may exhibit certain influence on the results of low temperature performance [33]. More accurate and reliable methods for the low temperature performance evaluation of PMA will be investigated in future studies [34].

Table 5. BBR test results.

Binder Type	−6 °C		−12 °C		−18 °C	
	Stiffness (MPa)	m-value	Stiffness (MPa)	m-value	Stiffness (MPa)	m-value
Pen60/70	N/A	N/A	280	0.280	541	0.199
CRM-0	84	0.362	172	0.355	330	0.205
CRM-A	62	0.388	117	0.332	235	0.277
CRM-B	58	0.406	103	0.347	213	0.268
CRM-C	62	0.395	109	0.339	256	0.247
SBR-0	107	0.339	203	0.308	347	0.192
SBR-A	89	0.356	188	0.322	352	0.186
SBR-B	84	0.350	180	0.310	308	0.220
SBR-C	80	0.352	173	0.324	316	0.215
SBS-0	97	0.360	175	0.317	316	0.208
SBS-A	66	0.423	143	0.342	275	0.255
SBS-B	74	0.426	155	0.359	289	0.238
SBS-C	70	0.440	140	0.336	260	0.265

3.6. Storage Stability

3.6.1. Softening Point Difference

Figure 7a shows the storage stability results of nano-montmorillonite-modified asphalt based on softening point difference. Smaller difference value (D-value) of softening points indicates superior storage stability. According to Stoke's law, the nano-montmorillonites tended to sink for its higher gravity (1.7–1.8) compared to virgin asphalt, regardless of its nanostructure, while the D-values of all NMAs met the requirement of AASHTO D5892, i.e., <2.5 °C. Additionally, the softening point difference values of VB-B and VB-C, which were smaller than 0.7 °C, exhibited an extremely stable dispersion of nano-montmorillonite in liquid asphalt. One possible reason may be that the colloidal size and the intercalated layer structure of nano-montmorillonite stopped the process of sinking. The colloidal size with adequate surface area can make the solid particles (nano-montmorillonite) move randomly in asphalt fractions rather than directly ascend (or descend) in vertical direction [35]. What's more, the stable disperse of nano-montmorillonites in virgin asphalt can also be attributed to the penetration of the asphalt fractions into the nano-montmorillonite layers, which modified the original structure to intercalated or exfoliated structure within asphalt components [17,21].

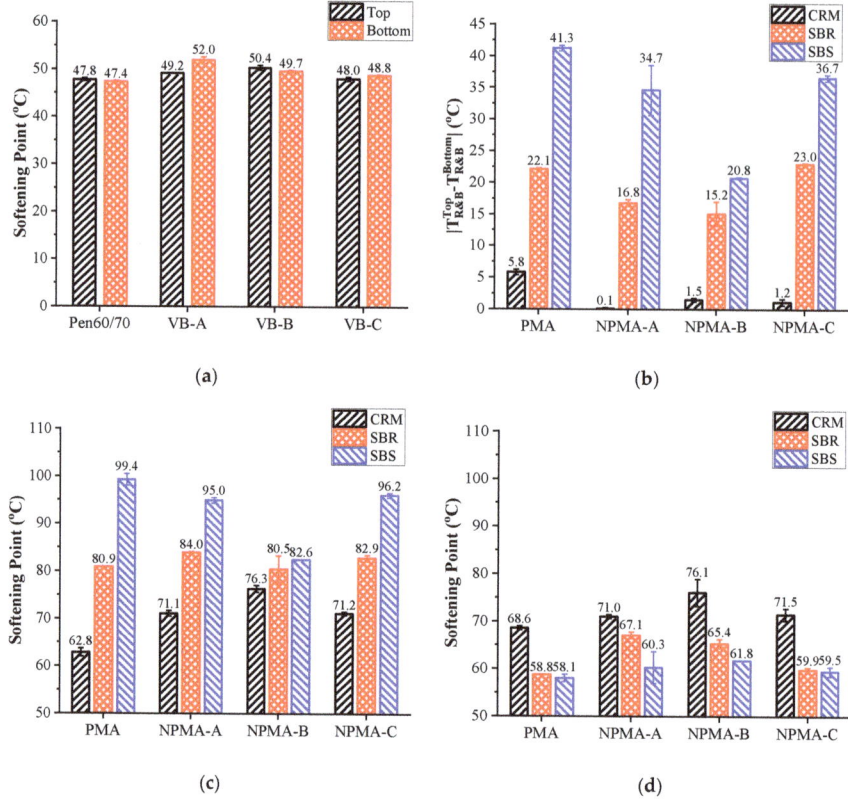

Figure 7. Softening point difference results: (**a**) Nano-montmorillonite modified asphalt; (**b**) D-values of NPMAs; (**c**) Softening points of top sections; (**d**) Softening points of bottom sections.

Figure 7b shows the D-values of modified binders, and Figure 7c,d present the softening point results of top and bottom sections after high-temperature storing, respectively. It is noted that the softening point difference value of SBR-0 was 16.3 °C higher than CRM-0, but 19.2 °C lower than SBS-0. According to the softening point results, this may be caused by the different modification effect of modifiers on the softening point, thus it is believed that all selected PMA binders had poor storage stability. For CRM and SBS modified binders, all three nano-montmorillonites decreased such differences. However, only Nanoclays A and B enhanced the storage stability of SBS-0. Comprehensively considering all test results, Nanoclay B seemed to work best on enhancing the storage stability of polymer-modified asphalt for the lowest D-values for SBR and SBS modified asphalt and the acceptable result for CRM-modified asphalt.

3.6.2. Complex Shear Modulus

According to previous studies [36,37], the complex shear modulus seemed to be a more reliable parameter to investigate the component variation of test binders. Therefore, the separation index (SI, Equation (2)) was proposed based on the complex modulus results by strategic highway research program (SHRP).

$$SI = \left(\frac{Max(G^*_{Top}, G^*_{Bottom}) - G^*_{Avg}}{G^*_{Avg}} \right) \times 100, \quad (2)$$

where G^*_{Top} and G^*_{Bottom} are the complex shear modulus at 25 °C at a frequency of 10 rad/s of the bottom and top parts after storage, and G^*_{Avg} is the average of G^*_{Top} and G^*_{Bottom}.

Table 6 presents the SI values at 25 °C (a typical intermediate temperature), 64 °C (a typical high temperature) and 82 °C (the rutting failure temperature for all PMA binders). It is noted that the SI value of SBR-0 was higher than CRM-0 but lower than SBS-0 at both intermediate and rutting failure temperatures, which is consistent with the results of softening point difference. After mixing with nano-montmorillonite, the *SI* value of CRM-0 at 25 °C and SBR-0 at 82 °C was significantly decreased, while there were no similar findings with SBS-modified binders. Therefore, a more efficient parameter was recommended. A best linear fit was firstly applied to obtain the temperature sensitivity parameter *k* (the slope of Equation (3)). Equation (2) was then used to calculate the separation index of *k*, i.e., SI_k. As shown in Table 7, it is noted that the storage stability of CRM-0 can be effectively enhanced by all selected nano-montmorillonites. Besides, the Nanoclay B worked well on all PMA binders, which was consistent with the softening point difference tests.

$$\log G^* = kT + b \qquad (3)$$

Table 6. Complex shear modulus results.

Binder Type	25 °C			64 °C			82 °C		
	Top (kPa)	Bottom (kPa)	SI (%)	Top (kPa)	Bottom (kPa)	SI (%)	Top (kPa)	Bottom (kPa)	SI (%)
CRM-0	1458.138	718.015	34.0	10.888	10.838	0.2	2.576	2.715	2.6
CRM-A	1065.133	1008.977	2.7	10.690	10.710	0.1	2.905	2.895	0.2
CRM-B	1062.840	961.900	5.0	14.545	15.166	2.1	4.143	4.242	1.2
CRM-C	886.238	970.832	4.6	11.611	12.159	2.3	3.102	3.370	4.1
SBR-0	464.028	3409.039	76.0	4.869	9.142	30.5	1.846	1.076	26.4
SBR-A	395.050	3039.114	77.0	4.456	9.686	37.0	1.731	1.489	7.5
SBR-B	366.635	2631.595	75.5	3.928	7.516	31.4	1.366	1.125	9.7
SBR-C	328.371	5072.979	87.8	5.790	12.139	35.4	2.477	1.516	24.1
SBS-0	160.889	1905.294	84.4	4.393	4.106	3.4	2.557	0.592	62.4
SBS-A	190.650	2231.642	84.3	7.319	5.218	16.8	4.791	0.882	68.9
SBS-B	145.732	2441.973	88.7	1.494	7.015	64.9	0.534	1.089	34.2
SBS-C	185.006	2945.014	88.2	8.393	7.067	8.6	5.774	0.972	71.2

Table 7. Temperature sensitivity evaluation results.

Modifiers	CRM			SBR			SBS		
	k_{Top}	k_{Bottom}	SI_k	k_{Top}	k_{Bottom}	SI_k	k_{Top}	k_{Bottom}	SI_k
N/A	−0.0493	−0.0432	6.6	−0.0435	−0.0621	17.6	−0.0329	−0.0626	31.1
Nanoclay A	−0.0460	−0.0456	0.5	−0.0428	−0.0590	16.0	−0.0294	−0.0610	34.9
Nanoclay B	−0.0432	−0.0421	1.2	−0.0439	−0.0601	15.6	−0.0441	−0.0598	15.2
Nanoclay C	−0.0439	−0.0441	0.2	−0.0385	−0.0627	23.9	−0.0277	−0.0621	38.3

3.6.3. Fourier Transform Infrared Spectroscopy

According to previous studies [38,39], the Fourier transform infrared (FTIR) tests can be used to evaluate the polymer content in modified asphalt. Infrared spectroscopy ranging from 4000 to 400 cm^{-1} was obtained by scanning using an FTIR spectrometer. In this study, the specific peak at 966 cm^{-1} (caused by out-of-plane bending γCH_3 vibration of trans-butadiene) was selected as a typical peak of CRM/SBR/SBS, while the peak at 1376 cm^{-1} (caused by in-plane bending vibration of δCH_3) was selected for virgin asphalt. Figure 8 shows the area of specific peak (Abs. 966 cm^{-1} and Abs.1376 cm^{-1}) under the FTIR curve. Then, the absorbance ratio (RA = Abs. 966 cm^{-1}/ Abs.1376 cm^{-1}) was calculated to present the CRM/SBR/SBS content. A larger absorbance ratio indicates a higher polymer content in the test sample.

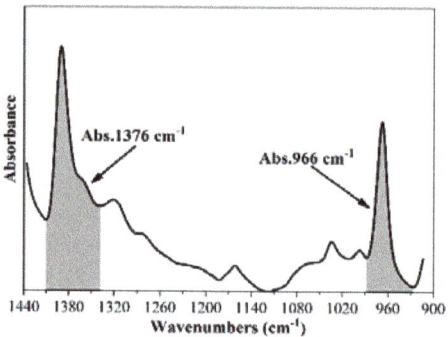

Figure 8. Calculation of area under specific peaks.

Table 8 present the FTIR results of the original binders and their corresponding top and bottom sections after storing. It is noted that the difference in RA values between top and bottom sections of CRM-0 was the largest among three PMA binders, which may indicate that the CRM-0 had the worst storage stability. Different from the other two tests, all types of nano-montmorillonites led to enhanced storage stability of the PMA binders. It can also be seen that nano-montmorillonites worked best on rubber-modified asphalt among three polymer modifiers, which was consistent with the softening point difference and complex shear modulus tests. What's more, Nanoclay B exhibited the best modification on storage stability among three nano-montmorillonites.

Table 8. FTIR test results.

Sample ID	Abs. 966 cm^{-1}			Abs. 1376 cm^{-1}			Absorbance Ratio (%)		
	Original Binder	Top	Bottom	Original Binder	Top	Bottom	Original Binder	Top	Bottom
CRM-0	0.554	10.406	1.084	0.021	0.044	0.100	3.746	0.424	9.226
CRM-A	0.731	1.377	0.428	0.047	0.077	0.012	6.384	5.595	2.870
CRM-B	0.815	1.687	1.238	0.042	0.072	0.061	5.199	4.273	4.909
CRM-C	0.541	0.733	1.316	0.032	0.037	0.065	5.869	5.110	4.971
SBR-0	6.415	2.625	2.690	0.496	0.360	0.095	7.739	13.701	3.530
SBR-A	4.597	4.921	3.350	0.307	0.492	0.176	6.675	9.990	5.265
SBR-B	3.729	6.747	5.551	0.326	0.674	0.319	8.745	9.988	5.747
SBR-C	6.189	7.049	2.572	0.596	1.065	0.317	9.625	15.109	12.318
SBS-0	2.392	4.730	9.357	0.381	1.949	0.716	15.938	41.201	7.650
SBS-A	3.377	3.740	6.663	0.566	1.478	0.529	16.773	39.522	7.939
SBS-B	3.907	3.656	8.985	0.518	1.323	1.341	17.231	36.193	14.926
SBS-C	1.506	4.512	8.391	0.300	1.929	1.102	20.327	42.762	13.132

3.7. Mechanism Investigation

The XRD tests were used to investigate the modification mechanism of nano-montmorillonites by determining their corresponding variation of layer distance when incorporated into polymer-modified asphalt. Based on the XRD analysis, the basal interlayer spacing (d) can be calculated from the first strong peak in the XRD spectra by means of the following equation:

$$2d \sin \theta = \lambda \tag{4}$$

Figure 9a shows d value of virgin asphalt, while the XRD results of nano-montmorillonites were presented in Figure 9b. It is noted that the d001 of virgin asphalt was quite small in the selected angle ranging from 1° to 10°. Among three nano-montmorillonites, Nanoclays A and B have only one peak while Nanoclay C has two peaks at different positions. The gap distance of Nanoclay C was the largest, while that of Nanoclay B was smallest. In mixed asphalt, the role of the nano-montmorillonite can be characterized according to the distance between the clay plates [23]. Specially, if the distance between the clay plates remain the same, the nano-montmorillonite act like regular particular fillers. In that case, the polymer cannot enter the layer structure of nano-montmorillonites. Conversely, an increased distance indicates the establishment of intercalated structures due to the penetration of polymer chains into the nano-montmorillonites.

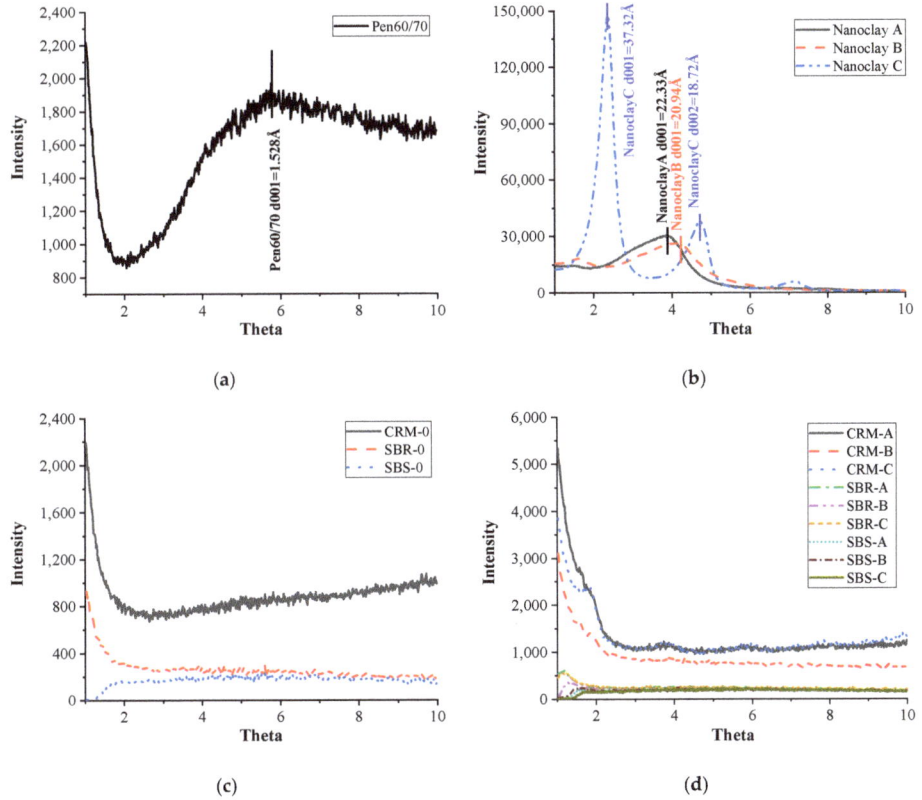

Figure 9. XRD test results: (**a**) Virgin asphalt; (**b**) Nano-montmorillonites; (**c**) Polymer-modified asphalt; (**d**) Nano-montmorillonite-polymer modified asphalt.

Figure 9c,d present the XRD results of PMA binders and NPMA results, respectively. It is noted that the clay interlayer diffraction peak can be observed at about 1°–2° in the results of NPMA binders, which shifted towards lower angles compared to the results of their corresponding nano-montmorillonites [24]. Nevertheless, there were no peaks noticed in the XRD spectra of PMA binders. Table 9 summarizes the d value results of nano-montmorillonite before and after being incorporated into modified asphalt. It is noted that the layer distance of nano-montmorillonite in the SBR-modified asphalt was the largest, while that in CRM-modified asphalt was the smallest regardless of the nano-montmorillonite type.

Table 9. Layer distance of nano-montmorillonite before and after mixed with asphalt.

Nano-Montmorillonite Type	Original Layer Distance (Å)	Measured Layer Distance in PMA (Increment) (Å)		
		CRM	SBR	SBS
Nanoclay A	22.33	46.49 (24.16)	73.28 (50.95)	53.48 (31.15)
Nanoclay B	20.09	48.42 (28.33)	65.41 (45.32)	53.90 (33.81)
Nanoclay C	37.32	48.33 (11.01)	71.71 (34.39)	50.29 (12.97)

By incorporating the colloid theory and the experimental results, the enhancement mechanism of nano-montmorillonites on storage stability can be explained as follows. Virgin asphalt is considered a dynamic colloidal system within which high molecular weight asphaltene micelles suspended in the lower molecular weight oily medium (maltenes) [40]. Since the layer distance of nano-montmorillonite did increase in the interaction process with polymers, the polymer chains penetrated the nanolayers were believed to belong to maltenes (in CRM-modified binders), SBR and SBS, respectively. The nano-montmorillonite structure became an intercalated structure rather than an exfoliated structure (Figure 10). With nano sizes and layer structures, the nano-montmorillonites do not settle down easily in hot asphalt even though their densities are larger than that of virgin asphalt. Additionally, with smaller layer distance change, the bonds between polymers and nano-montmorillonite were considered stronger due to Van der Waals forces between nano-montmorillonite layers. Eventually, the existence of nano-layers slowed down the precipitation of polymer modifiers, which results in a more stable disperse of the modifier in virgin asphalt, therefore improving the storage stability of PMA binders.

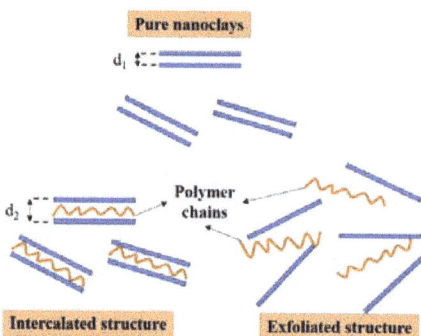

Figure 10. Schematic illustration of intercalated and exfoliated structure.

4. Conclusions

This study evaluated the effect of nano-montmorillonite on the properties of modified asphalt with different polymers including CRM, SBR and SBS. Rheological and chemical tests were conducted to characterize the rheological properties of obtained NPMA binders and reveal the modification mechanism of nano-montmorillonite on the storing performance. According to test results, the findings are obtained as follows.

- Adding nano-montmorillonite slightly improved the viscosity of PMA binders, and had insignificant effects on their rheological properties.
- The incorporation of all three types of nano-montmorillonites effectively alleviates the storage stability concern of CRM-0, while having a limited positive effect on SBR-0 and SBS-0.
- Nanoclay B, pure montmorillonite with Hydroxyl organic ammonium, exhibited the most obvious effect in improving the storage stability of the three selected PMA binders.

This study suggests that nano-montmorillonite is a promising modifier to alleviate the storage stability concern for asphalt with polymer modifiers. Future studies will focus on investigating the interaction mechanism among nano-montmorillonite, virgin asphalt and different modifiers, as well as the effects of nano-montmorillonites on the engineering performance of asphalt mixture.

Author Contributions: Data curation, Z.R. Funding acquisition, H.Y. and J.Y. Investigation, Y.Z., F.G., Z.R., M.Z., and H.Y. Methodology, Z.R., and H.Y. Supervision, H.L., Writing—original draft, Z.R. Writing—review & editing, Q.W., and H.Y. All authors have read and agreed to the published version of the manuscript.

Funding: The National Natural Science Foundation of China [NSFC 51808228 and 51678251], The open funding of Key Laboratory for Special Area Highway Engineering [300102219517] and the open funding of Research and Development Center of Transport Industry of Technologies, Materials and Equipment of Highway Construction and Maintenance, Gansu Road & Bridge Group [No. GLKF201801].

Conflicts of Interest: The authors confirm that there are no conflicts of interest associated with this manuscript.

References

1. Wang, D.; Falchetto, A.C.; Poulikakos, L.; Hofko, B.; Porot, L. RILEM TC 252-CMB report: Rheological modeling of asphalt binder under different short and long-term aging temperatures. *Mater. Struct.* **2019**, *52*, 73. [CrossRef]
2. Behnke, R.; Wollny, I.; Hartung, F.; Kaliske, M. Thermo-Mechanical Finite Element Prediction of the Structural Long-Term Response of Asphalt Pavements Subjected to Periodic Traffic Load: Tire-Pavement Interaction and Rutting. *Comput. Struct.* **2019**, *218*, 9–31. [CrossRef]
3. Yang, Z.; Zhang, X.; Zhang, Z.; Zou, B.; Zhu, Z.; Lu, G.; Yu, H. Effect of aging on chemical and rheological properties of bitumen. *Polymers* **2018**, *10*, 1345. [CrossRef] [PubMed]
4. Fang, H.; Zou, F.; Liu, W.; Wu, C.; Bai, Y.; Hui, D. Mechanical Performance of Concrete Pavement Reinforced by CFRP Grids for Bridge Deck Applications. *Compos. Part B Eng.* **2017**, *110*, 315–335. [CrossRef]
5. Yu, H.; Leng, Z.; Dong, Z.; Tan, Z.; Guo, F.; Yan, J. Workability and mechanical property characterization of asphalt rubber mixtures modified with various warm mix asphalt additives. *Constr. Build. Mater.* **2018**, *175*, 392–401. [CrossRef]
6. Xie, J.; Yang, Y.; Lv, S.; Zhang, Y.; Zhu, X.; Zheng, C. Investigation on Rheological Properties and Storage Stability of Modified Asphalt Based on the Grafting Activation of Crumb Rubber. *Polymers* **2019**, *11*, 1563. [CrossRef]
7. Yu, H.; Zhu, Z.; Leng, Z.; Wu, C.; Zhang, Z.; Wang, D.; Oeser, M. Effect of mixing sequence on asphalt mixtures containing waste tire rubber and warm mix surfactants. *J. Clean. Prod.* **2020**, *246*, 119008. [CrossRef]
8. Yu, H.; Leng, Z.; Zhang, Z.; Li, D.; Zhang, J. Selective absorption of swelling rubber in hot and warm asphalt binder fractions. *Constr. Build. Mater.* **2020**, *238*, 117727. [CrossRef]
9. Miao, Y.; Wang, T.; Wang, L. Influences of Interface Properties on the Performance of Fiber-Reinforced Asphalt Binder. *Polymers* **2019**, *11*, 542. [CrossRef]
10. Ghavibazoo, A.; Abdelrahman, M.; Ragab, M. Effect of Crumb Rubber Modifier Dissolution on Storage Stability of Crumb Rubber-Modified Asphalt. *Trans. Res. Rec.* **2013**, *2370*, 109–115. [CrossRef]
11. Zanzotto, L.; Kennepohl, G.J. Development of Rubber and Asphalt Binders by Depolymerization and Devulcanization of Scrap Tires in Asphalt. *Trans. Res. Rec.* **1996**, *91*, 32–38. [CrossRef]
12. Golestani, B.; Nam, B.H.; Nejad, F.M.; Fallah, S. Nanoclay Application to Asphalt Concrete: Characterization of Polymer and Linear Nanocomposite-Modified Asphalt Binder and Mixture. *Constr. Build. Mater.* **2015**, *91*, 32–38. [CrossRef]
13. Fini, E.H.; Hosseinnezhad, S.; Oldham, D.; Mclaughlin, Z.; Alavi, Z.; Harvey, J. Bio-modification of rubberised asphalt binder to enhance its performance. *Int. J. Pavement Eng.* **2019**, *20*, 1216–1225. [CrossRef]
14. Yu, J.; Ren, Z.; Gao, Z.; Wu, Q.; Zhu, Z.; Yu, H. Recycled Heavy Bio Oil as Performance Enhancer for Rubberized Bituminous Binders. *Polymers* **2019**, *11*, 800. [CrossRef]
15. Cheng, G.; Shen, B.; Zhang, J. A Study on the Performance and Storage Stability of Crumb Rubber-Modified Asphalts. *Petrol. Sci. Tech.* **2011**, *29*, 192–200. [CrossRef]
16. Yu, R.; Fang, C.; Liu, P.; Liu, X.; Li, Y. Storage Stability and Rheological Properties of Asphalt Modified with Waste Packaging Polyethylene and Organic Montmorillonite. *Appl. Clay Sci.* **2015**, *104*, 1–7. [CrossRef]

17. Daniel Martinez-Anzures, J.; Zapien-Castillo, S.; Adriana Salazar-Cruz, B.; Luis Rivera-Armenta, J.; Del Carmen Antonio-Cruz, R.; Hernandez-Zamora, G.; Leonor Mendez-Hernandez, M. Preparation and Properties of Modified Asphalt Using Branch SBS/Nanoclay Nanocomposite as a Modifier. *Road Mater. Pavement Des.* **2019**, *20*, 1275–1290. [CrossRef]
18. Vargas, M.A.; Moreno, L.; Montiel, R.; Manero, O.; Vazquez, H. Effects of Montmorillonite (Mt) and Two Different Organo-Mt Additives on the Performance of Asphalt. *Appl. Clay Sci.* **2017**, *139*, 20–27. [CrossRef]
19. Fang, C.; Yu, R.; Liu, S.; Li, Y. Nanomaterials Applied in Asphalt Modification: A Review. *J. Mater. Sci. Tech.* **2013**, *29*, 589–594. [CrossRef]
20. Guo, S.; Dai, Q.; Wang, Z.; Yao, H. Rapid Microwave Irradiation Synthesis of Carbon Nanotubes on Graphite Surface and Its Application on Asphalt Reinforcement. *Compos. Part B Eng.* **2017**, *124*, 134–143. [CrossRef]
21. Yu, J.; Ren, Z.; Yu, H.; Wang, D.; Svetlana, S.; Korolev, E.; Gao, Z.; Guo, F. Modification of Asphalt Rubber with Nanoclay towards Enhanced Storage Stability. *Materials* **2018**, *11*, 2093. [CrossRef] [PubMed]
22. Leng, Z.; Tan, Z.; Yu, H.; Guo, J. Improvement of storage stability of SBS-modified asphalt with nanoclay using a new mixing method. *Road Mater. Pavement Des.* **2019**, *20*, 1601–1614. [CrossRef]
23. Galooyak, S.S.; Dabir, B.; Nazarbeygi, A.E.; Moeini, A. Rheological Properties and Storage Stability of Bitumen/SBS/Montmorillonite Composites. *Constr. Build. Mater.* **2010**, *24*, 300–307. [CrossRef]
24. Abdelrahman, M.; Katti, D.R.; Ghavibazoo, A.; Upadhyay, H.B.; Katti, K.S. Engineering Physical Properties of Asphalt Binders through Nanoclay-Asphalt Interactions. *J. Mater. Civ. Eng.* **2014**, *26*, 04014099. [CrossRef]
25. Navarro, F.J.; Partal, P.; Martinez-Boza, F.; Gallegos, C. Thermo-Rheological Behaviour and Storage Stability of Ground Tire Rubber-Modified Bitumens. *Fuel* **2004**, *83*, 2041–2049. [CrossRef]
26. Ghavibazoo, A.; Abdelrahman, M. Composition Analysis of Crumb Rubber During Interaction with Asphalt and Effect on Properties of Binder. *Int. J. Pavement Eng.* **2013**, *14*, 517–530. [CrossRef]
27. ASTM Standard D7173. *Standard Practice for Determining the Separation Tendency of Polymer from Polymer Modified Asphalt*; American Society for Testing and Materials: West Conshohocken, PA, USA, 2014.
28. Yu, H.; Zhu, Z.; Zhang, Z.; Yu, J.; Oeser, M.; Wang, D. Recycling Waste Packaging Tape into Bituminous Mixtures Towards Enhanced Mechanical Properties and Environmental Benefits. *J. Clean. Prod.* **2019**, *229*, 22–31. [CrossRef]
29. Bairgi, B.K.; Mannan, U.A.; Tarefder, R.A. Tribological Evaluation for an in-Depth Understanding of Improved Workability of Foamed Asphalt. *Trans. Res. Rec.* **2019**, *2673*, 533–545. [CrossRef]
30. Yu, H.; Leng, Z.; Zhou, Z.; Shih, K.; Xiao, F.; Gao, Z. Optimization of Preparation Procedure of Liquid Warm Mix Additive Modified Asphalt Rubber. *J. Clean. Prod.* **2017**, *141*, 336–345. [CrossRef]
31. Yu, H.; Zhu, Z.; Wang, D. Evaluation and Validation of Fatigue Testing Methods for Rubberized Bituminous Specimens. *Trans. Res. Rec.* **2019**, *2673*, 603–610. [CrossRef]
32. Zhou, F.; Mogawer, W.; Li, H.; Andriescu, A.; Copeland, A. Evaluation of Fatigue Tests for Characterizing Asphalt Binders. *J. Mater. Civ. Eng.* **2013**, *25*, 610–617. [CrossRef]
33. Wang, D.; Falchetto, A.C.; Riccardi, C.; Westerhoff, J.; Wistuba, M.P. Investigation on the effect of physical hardening and aging temperature on low-temperature rheological properties of asphalt binder. *Road Mater. Pavement Des.* **2019**, 1–23. [CrossRef]
34. Wang, D.; Falchetto, A.C.; Alisov, A.; Schrader, J.; Riccardi, C.; Wistuba, M.P. An Alternative Experimental Method for Measuring the Low Temperature Rheological Properties of Asphalt Binder by Using 4 mm Parallel Plates on Dynamic Shear Rheometer. *Trans. Res. Rec.* **2019**, *2673*, 427–438.
35. Hwang, Y.; Lee, J.-K.; Lee, J.-K.; Jeong, Y.-M.; Cheong, S.-I.; Ahn, Y.-C.; Kim, S.H. Production and Dispersion Stability of Nanoparticles in Nanofluids. *Powder Tech.* **2008**, *186*, 145–153. [CrossRef]
36. Xie, J.; Yang, Y.; Lv, S.; Peng, X.; Zhang, Y. Investigation on Preparation Process and Storage Stability of Modified Asphalt Binder by Grafting Activated Crumb Rubber. *Materials* **2019**, *12*, 2014. [CrossRef]
37. Nasr, D.; Pakshir, A.H. Rheology and Storage Stability of Modified Binders with Waste Polymers Composites. *Road Mater. Pavement Des.* **2019**, *20*, 773–792. [CrossRef]
38. Mouillet, V.; Lamontagne, J.; Durrieu, F.; Planche, J.-P.; Lapalu, L. Infrared Microscopy Investigation of Oxidation and Phase Evolution in Bitumen Modified with Polymers. *Fuel* **2008**, *87*, 1270–1280. [CrossRef]

39. Masson, J.F.; Pelletier, L.; Collins, P. Rapid FTIR Method for Quantification of Styrene-Butadiene Type Copolymers in Bitumen. *J. Appl. Poly. Sci.* **2001**, *79*, 1034–1041. [CrossRef]
40. Loeber, L.; Muller, G.; Morel, J.; Sutton, O. Bitumen in Colloid Science: A Chemical, Structural and Rheological Approach. *Fuel* **1998**, *77*, 1443–1450. [CrossRef]

© 2020 by the authors. Licensee MDPI, Basel, Switzerland. This article is an open access article distributed under the terms and conditions of the Creative Commons Attribution (CC BY) license (http://creativecommons.org/licenses/by/4.0/).

Article

Effect of Borpolymer on Mechanical and Structural Parameters of Ultra-High Molecular Weight Polyethylene

Sakhayana N. Danilova [1,*], Afanasy A. Dyakonov [1,2], Andrey P. Vasilev [1], Aitalina A. Okhlopkova [1], Aleksei G. Tuisov [3], Anatoly K. Kychkin [2] and Aisen A. Kychkin [3]

1 Department of Chemistry, Institute of Natural Sciences, North-Eastern Federal University, 677013 Yakutsk, Russia; afonya71185@mail.ru (A.A.D.); gtvap@mail.ru (A.P.V.); okhlopkova@ya.ru (A.A.O.)
2 Institute of the Physical-Technical Problems of the North, Siberian Branch of the Russian Academy of Sciences, 677980 Yakutsk, Russia; kychkinplasma@mail.ru
3 Federal Research Centre "The Yakut Scientific Centre of the Siberian Branch of the Russian Academy of Sciences", 677000 Yakutsk, Russia; tuisovag@gmail.com (A.G.T.); icen.kychkin@mail.ru (A.A.K.)
* Correspondence: dsn.sakhayana@mail.ru

Abstract: The paper presents the results of studying the effect of borpolymer (BP) on the mechanical properties, structure, and thermodynamic parameters of ultra-high molecular weight polyethylene (UHMWPE). Changes in the mechanical characteristics of polymer composites material (PCM) are confirmed and complemented by structural studies. X-ray crystallography (XRC), differential scanning calorimetry (DSC), scanning electron microscopy (SEM), and infrared spectroscopy (IR) were used to study the melting point, morphology and composition of the filler, which corresponds to the composition and data of the certificate of the synthesized BP. Tensile and compressive mechanical tests were carried out in accordance with generally accepted standards (ASTM). It is shown that BP is an effective modifier for UHMWPE, contributing to a significant increase in the deformation and strength characteristics of the composite: tensile strength of PCM by 56%, elongation at break by 28% and compressive strength at 10% strain by 65% compared to the initial UHMWPE, due to intensive changes in the supramolecular structure of the matrix. Structural studies revealed that BP does not chemically interact with UHMWPE, but due to its high adhesion to the polymer, it acts as a reinforcing filler. SEM was used to establish the formation of a spherulite supramolecular structure of polymer composites.

Keywords: ultra-high molecular weight polyethylene; polymethylene-p-triphenyl ester of boric acid; borpolymer; polymer composite materials

1. Introduction

Currently, polymer composite materials (PCM), due to their high mechanical properties and other special characteristics, low density, and ease of industrial processing, are widely used in industry, medicine, and other fields. Ultra-high molecular weight polyethylene (UHMWPE) is one of the promising polymers for manufacturing structural PCMs. It is known that UHMWPE is characterized by high chemical inertness, excellent mechanical properties, high impact strength and low coefficient of friction [1]. Due to these properties, UHMWPE is used, and can potentially be used in many areas from medicine to the space industry [2,3]. The introduction of micro- and nanosized fillers into UHMWPE increases the mechanical and tribological characteristics, which expands the application range of the material [4,5]. It is known that polymer composites based on UHMWPE filled with nanosized fillers are distinguished by a low coefficient of friction, increased strength characteristics, and resistance to cracking [6,7]. Due to the high specific surface area of particles and the decompensation of bonds of a significant number of atoms, nanosized fillers are characterized by their agglomeration, which leads to the appearance of defective regions and, consequently, to a decrease in the mechanical characteristics of PCM. There are studies on the modification of UHMWPE by the introduction of fibrous fillers [8,9], where there is an increase in the bearing capacity, wear resistance, rigidity, and strength of PCM [4]. The mechanical characteristics of fiber-filled composites depend on the interfacial interaction at the "fiber-polymer" interface, which requires additional modification of the fiber surface or the introduction of adhesion promoters into the PCM composition [10–12]. There are investigations in which polymers are used as a filler for UHMWPE. Such materials are characterized by increased wear resistance and low coefficient of friction [13–16], but at the same time they have low deformation and strength characteristics. For example, in the case such organic fillers as polyetheretherketones (PEEK) when creating composites based on UHMWPE, a decrease in mechanical parameters is shown [16]. A great number of studies are devoted to UHMWPE/PEEK composites and their wide application in the development of bone and hip implants, due to the biological characteristics of PEEK [17–20]. In [20], it was found that PEEK is poorly compatible with UHMWPE. However, a slight increase in the mechanical parameters of composites is explained by the high hardness of PEEK particles in comparison with UHMWPE [16].

In addition to the use of fillers to improve UHMWPE properties, inoculation techniques are used, including methods of ultrasonic treatment, mechanical activation, high-speed mixing of composite components, etc. [21–23]. In addition, specific methods of processing PCM are used, including the following: mixing a filler and polymer in solvents, adding surfactants [24–26], crosslinking UHMWPE macromolecules, modifying fillers by CVD—chemical vapor deposition [27], functionalizing fillers, etc. [28,29]. Despite a large number of publications on the study of the modification of UHMWPE composites components, the mechanisms for realizing the potential capabilities of PCM components have not yet been disclosed.

In this study, we investigated the effect of polymethylene-p-triphenyl ester of boric acid (BP) on the mechanical properties, structure, and thermodynamic parameters of UHMWPE, depending on its content. Polymethylene-p-triphenyl ester of boric acid is a class of organic boron compounds in which the B (boron) atom in the phenol molecule is linked through the O (oxygen) atom. Organic boron compounds are widely used in various fields: to increase the fire resistance of materials [30], to obtain porous materials [31], and as a polymer modifier [32–34]. There is a great deal of studies devoted to BP as an additive in epoxy resin and rubber [35–40]. However, borpolymers, in particular BP, as a filler for UHMWPE have not been investigated practically.

The aim of this research is to study the effect of boron polymer on the mechanical properties and structure of ultra-high molecular weight polyethylene.

2. Materials and Methods

2.1. Materials and Obtaining of PCM

UHMWPE brand GUR-4022 (Celanese, Nanjing, China) was used as a polymer matrix, with a molecular weight of 5.0×10^6 g/mol, a density of 0.93 g/cm^3, and an average particle size of 145 µm. A synthesized polymethylene-p-triphenyl boric acid ester-PTBEC, called borpolymer (BP), was used as a modifying additive. BP was provided by Boroplast LLC (Boroplast, Biysk, Russia), with an average molecular weight of 2500–3000 a.u. and with a melting point of 150–160 °C.

To remove adsorbed moisture, the initial UHMWPE powder was preliminarily dried in a PE-0041 oven (Ekopribor, St. Petersburg, Russia) at a temperature of 85 °C for 1.5 h. UHMWPE and BP powder were mixed at room temperature in a paddle mixer with a rotor speed of 1200 rpm. The samples were prepared using the hot pressing technology in a PCMV-100 hydraulic vulcanization press (Impulse, Ivanovo, Russia) at a temperature of 175 °C, a pressure of 10 MPa, holding for 20 min and then cooling to room temperature. The borpolymer content in the polymer matrix was varied: 0.2, 0.3, 0.5, 1.0, 2.0, 3.0, and 5.0 wt. %.

2.2. Research Methods

The mechanical properties of UHMWPE and PCM were studied using the Autograph AGS-J tensile testing machine (Shimadzu, Tokyo, Japan). The tensile strength and elongation at break were tested according to ASTM D3039/D3039M-14 at the moving gripper speed of 50 mm/min, the number of samples was six. Compressive strength was determined according to ASTM D695.

X-ray diffraction patterns of the borpolymer and PCM was determine using X-ray powder diffractometry (XRD, ARL X'Tra, Thermo Fisher Scientific, Ecublens, Switzerland). An X-ray tube with a copper anode (λ (CuK$_\alpha$) = 0.154 nm) was used as a radiation source. For the study, we used samples in the form of plates with dimensions of 30 × 30 × 3 mm. The degree of crystallinity was determined by the Formula (1):

$$\alpha = \frac{A_c}{A_c + A_a} * 100\%, \tag{1}$$

where A_c is the area under the crystalline peaks, and $A_c + A_a$ is the total area of both crystalline and amorphous regions. The average crystallite size (L) in the direction perpendicular to the crystal lattice plane was determined using the Scherrer Equation (2):

$$L = \frac{K\lambda}{\beta \cos \theta}, \tag{2}$$

where β is the width at half maximum of the diffraction peak; K is the crystal lattice constant (approximately 0.9); λ—wavelength of the beam of monochromatic radiation CuK$_\alpha$, 0.154 nm; θ corresponds to the Bragg angle, and L corresponds to the average crystallite size. The distance (d) between the diffraction planes was obtained according to Bragg's law (3):

$$2d \sin \theta = n\lambda, \tag{3}$$

where n is the diffraction order (integer); d—interplanar distance; λ is the wavelength of X-ray radiation, θ is the Bragg angle.

The supramolecular structure of UHMWPE and PCM and powders of BP were studied on the JSM-7800F scanning electron microscope (Jeol, Akishima, Japan) with the X-MAX-20 attachment (Oxford Instruments plc, Tubney Woods, Abingdon, UK) in the secondary electron mode at an accelerating voltage of 1–1.5 kV.

The atomic force microscopy was performed with an NTEGRA instrument manufactured by NTegra Prima (NT-MTD, Zelenograd, Russia). The instrument was operated in 'semi-contact mode', which is often also referred to as "tapping mode". Surface topography and phase images were obtained using NSG 10 golden silicone probes with a resonant frequency of 140–390 kHz and a force constant of 2.5–10 N/m. The AFM images obtained were processed using the "Nova" and "Image Analysis" software (NT-MTD, Zelenograd, Russia).

Fourier transform infrared (IR) spectroscopy (FTIR; Varian 7000, Palo Alto, CA, USA) was used to record IR spectra with an attenuated total reflection (ATR) attachment over the range 400–4000 cm^{-1}.

The Raman spectra between 1600 cm^{-1} and 1000 cm^{-1} were recorded by using the NT-MDT NTEGRA (NT-MDT, Zelenograd, Russia) equipped in a 532 nm. The spectra were collected on three different points in one sample.

The thermodynamic characteristics of UHMWPE and composites were studied on a DSC 204 F1 Phoenix NETZSCH differential scanning calorimeter (Netzsch, Selb, Germany), where the measurement error was not more than ±0.1%, the heating rate was 20 °C/min, and the sample weight was 18 ± 1 mg. The measurements were carried out in a helium medium in a temperature range of 40–180 °C. The samples were placed in aluminum crucibles with a 40 µL. Temperature calibration was performed using standard samples of In, Sn, Bi, Pb, and KNO$_3$.

The degree of crystallinity of UHMWPE and PCM was calculated by the follow Equation (4):

$$\alpha, \% = \frac{\Delta H_{endotherm}}{\Delta H_f \left(1 - W_f\right)} \cdot 100\%, \qquad (4)$$

where $\Delta H_{endotherm}$—is the melting enthalpy calculated from the area of endothermic melting peak; ΔH_f—is the melting enthalpy for 100% crystalline UHMWPE, which is equal to 291 J/g; W_f—is the mass content of the filler in PCM [1,41].

3. Results and Discussion

3.1. Characteristics of Borpolymer

Borpolymer belongs to the class of boric esters with a molecular weight distribution of the basic substance ≥99%. This substance is obtained by the polycondensation reaction of triphenyl ester of boric acid and 1, 3, 5—trioxane (paraformaldehyde) in an acidic medium.

Figure 1 shows the X-ray diffractogram of borpolymer. Based on the analysis of XRC diffraction patterns, it was established that the initial borpolymer is an amorphous compound, and a broad peak characteristic of amorphous compounds with a low intensity in the region (2θ = 10–30°) was found. No other peaks in the study area 2θ = 1.5–60° were found in the tested BP sample.

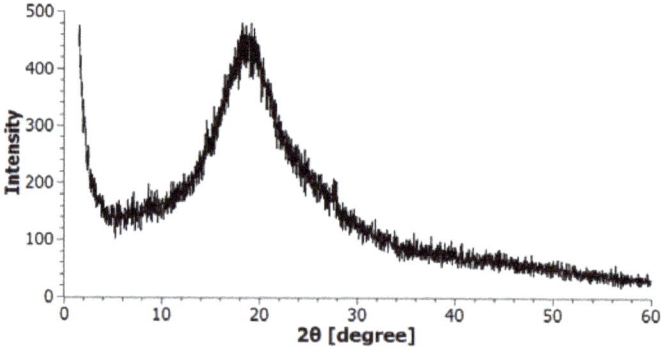

Figure 1. XRD pattern of raw BP.

For a qualitative analysis of the BP composition, the BP structure was studied by IR spectroscopy (Figure 2).

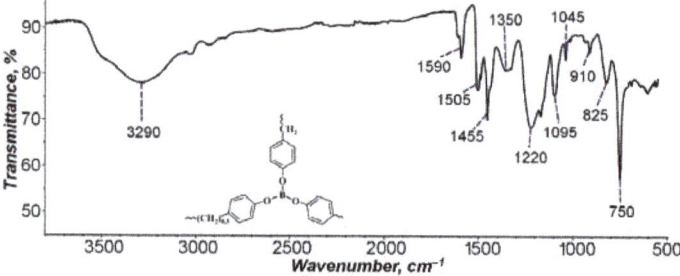

Figure 2. IR spectra of borpolymer.

As Figure 2 demonstrates, the IR spectrum shows the following characteristic peaks of BP benzene rings: 1045 and 1095 cm^{-1}, corresponding to the vibrations of the C–H bond (methyl radical) in the plane of the benzene ring (in plane C–H blending), and a peak at 750 cm^{-1} due to vibration outside the plane of the benzene ring of the C–H bond of the methyl group (out of plane C–H blending). The peaks at 1590–1455 cm^{-1} correspond to the vibrations of the C=C bonds of the aromatic ring itself, and the intense absorption band in the region at 3290 cm^{-1} refers to the vibrations of the C–H bonds of the benzene ring [42]. The peaks of absorption bands of carbon-boron and oxygen-boron bonds were also found. Asymmetric stretching vibrations of the B–C bond in triphenylboron correspond to a peak at 1220 cm^{-1}, while symmetric vibrations of the B–C bond are characterized by the occurrence of a peak at 825 cm^{-1}. The peak at 1350 cm^{-1} is caused by bending vibrations of the B–O bond. Symmetric vibrations of this bond are marked by the occurrence of a low-intensity peak at 910 cm^{-1}; possibly, the intensity decreases due to the strength of the B–O bond in the BP polymer [43]. The obtained IR spectra correspond to the chemical composition of BP.

Analysis of the DSC data indicates the presence of two melting peaks at 74 °C and 150 °C. The presence of two peaks indicates a polydisperse molecular weight distribution of the borpolymer, as the lower molecular weight portion of BP begins to melt at a relatively low temperature. The main melting peak on the DSC BP curve corresponds to 150 °C, and with its increase BP completely transforms into a molten state. It is known that BP is one of the promising heat-resistant additives for thermosetting plastics that increase the strength and wear resistance of materials [35–38]. Based on the temperature data, BP is suitable for the processing temperature range of UHMWPE based composites.

The sizes and morphology of the crushed BP particles were studied using a scanning electron microscope, the micrographs of which are shown in Figure 3.

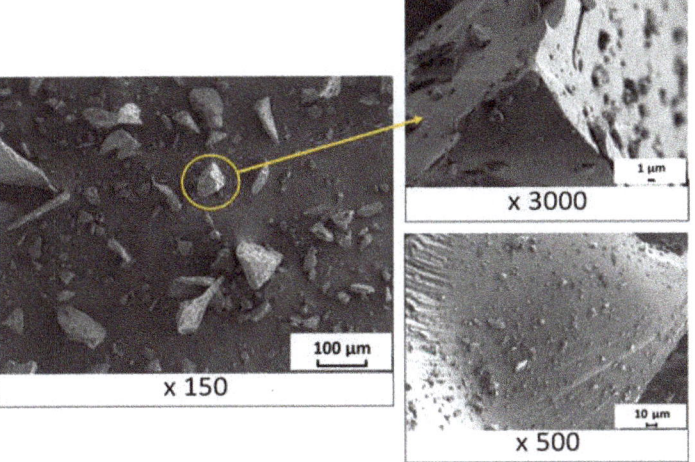

Figure 3. Micrographs of BP particles.

The micrographs in Figure 3 show that the surface of the crushed particles is characterized by microdefects resulting from brittle fracture of BP. It was found that glass-like particles of BP are easily crushed; nevertheless, there is a wide variation in the size of crushed particles. It is noteworthy that the smaller BP particles are deposited on the surface of the larger ones. Obviously, at the stage of mixing the components of the polymer composition in a paddle mixer, due to mechanical effects, the BP particles will be dispersed with a fairly uniform distribution in the volume of the polymer.

3.2. Study of the PCM Structure
3.2.1. IR Spectra of Composites

In order to determine the chemical effect of BP on the polymer matrix, the IR spectra of the initial UHMWPE and the UHMWPE/5 wt. % BP composite were studied (Figure 4).

The IR spectra revealed the main peaks of UHMWPE at 2920, 2850, and 1470 cm^{-1}, related to stretching and bending vibrations of -CH$_2$ bonds and 1365 cm^{-1}, corresponding to bending vibrations of -CH$_3$. A crystallinity peak at 720 cm^{-1} was also found, due to pendulum vibrations of the polymer chain.

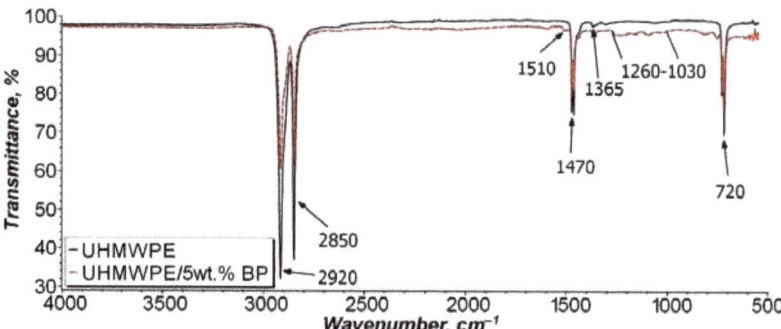

Figure 4. IR spectra of the initial UHMWPE and composite UHMWPE/5 wt. % BP.

As can be seen from Figure 4, the IR spectrum of the UHMWPE/5 wt. % BP composite differs from the initial UHMWPE by the appearance of a broad absorption band in the 1260–1030 cm^{-1} region, which is characteristic of vibrations of the boron—oxygen bond, ether bond, or methyl group relative to the plane of the benzene ring. In addition, there is an insignificant peak at 1510 cm^{-1}, which indicates the presence of carboxyl groups (C=O) or a C=C bond of the benzene ring of BP. It can be seen that the intensity of the detected peaks is minimal. It can be assumed that BP will not chemically interact with UHMWPE macromolecules and will not oxidize during PCM processing.

3.2.2. Morphology of PCM

During the formation of composites and at the stages of processing, structural changes occur associated with a change in the supramolecular structure and the development of molecular orientation of polymer macromolecules. These changes in the structure of the polymer matrix determine the complex of mechanical properties of PCM. Due to the difference in the formation of the supramolecular structure, composites of the same polymer are often characterized by different values of mechanical parameters. Therefore, in order to determine the processes of structure formation in the supramolecular structure, UHMWPE and PCM conducted a SEM study, the results of which are shown in Figure 5.

Figure 5 shows that the supramolecular structure of the initial UHMWPE is characterized by a lamellar structure. The introduction of BP into UHMWPE transforms the lamellar structure into a spherulite structure of the radial type with irregularly shaped elements. Composite with 0.5 wt. % BP is characterized by the formation of large spherulites with clearly defined boundaries. In the case of 1 wt. % BP in UHMWPE, a decrease in the size of spherulite structures is observed. The supramolecular structure of the composite containing 2 and 5 wt. % filler becomes more disordered, defect regions are recorded, which will further affect the mechanical properties of the material. At the same time, these composites contain fan-shaped spherulites.

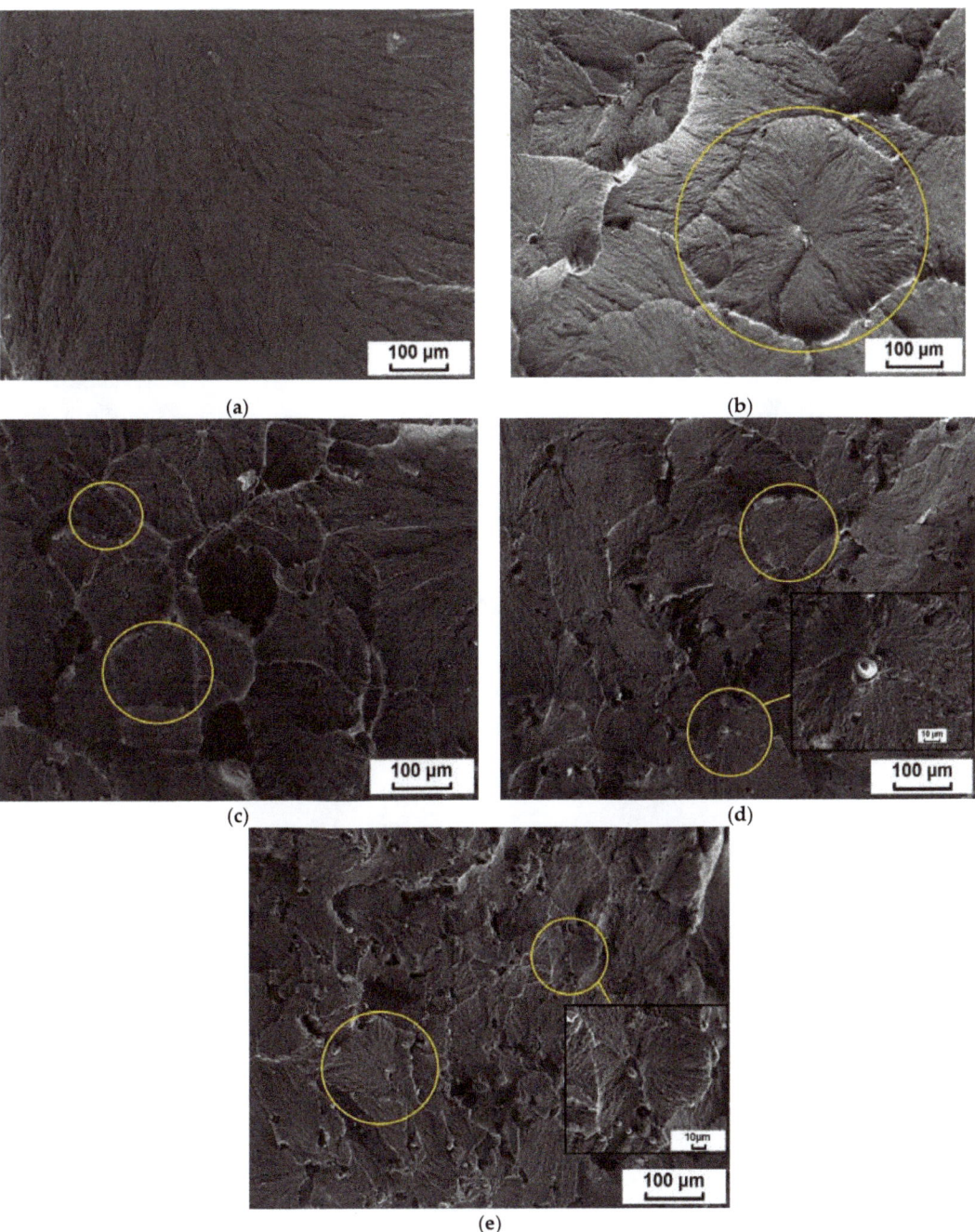

Figure 5. Microphotographs of the structure of (**a**) the initial ultra-high-molecular weight polyethylene (UHMWPE) and polymer composite materials (PCMs) based on UHMWPE filed by BP (**b**) 0.5 wt. %, (**c**) 1.0 wt. %, (**d**) 2.0 wt. % and (**e**) 5.0 wt. %.

3.2.3. Investigation of the Structure of Composites by the AFM Method

Structural studies of the composites were carried out using the AFM method in a semicontact mode, which makes it possible to obtain a high contrast in the visualization of submicron structures and to recognize various components in heterogeneous polymer systems (Figure 6). In the case of a smooth but chemically dissimilar surface, it is possible to visualize surface areas that differ in phase composition. Since the detection of the oscillation phase occurs simultaneously with the acquisition of the surface topography with the amplitude detection of the probe position in the feedback, it is possible to obtain information on the phase composition of the sample from the comparison of the amplitude and phase images. In this work, the object of study was a composite based on UHMWPE and BP, where the latter particles act as a dispersed phase. Therefore, the phase-contrast on the AFM made it possible to estimate the degree of BP distribution in the volume of the matrix and to measure the size of the crushed filler particles during the processing of the composite [44].

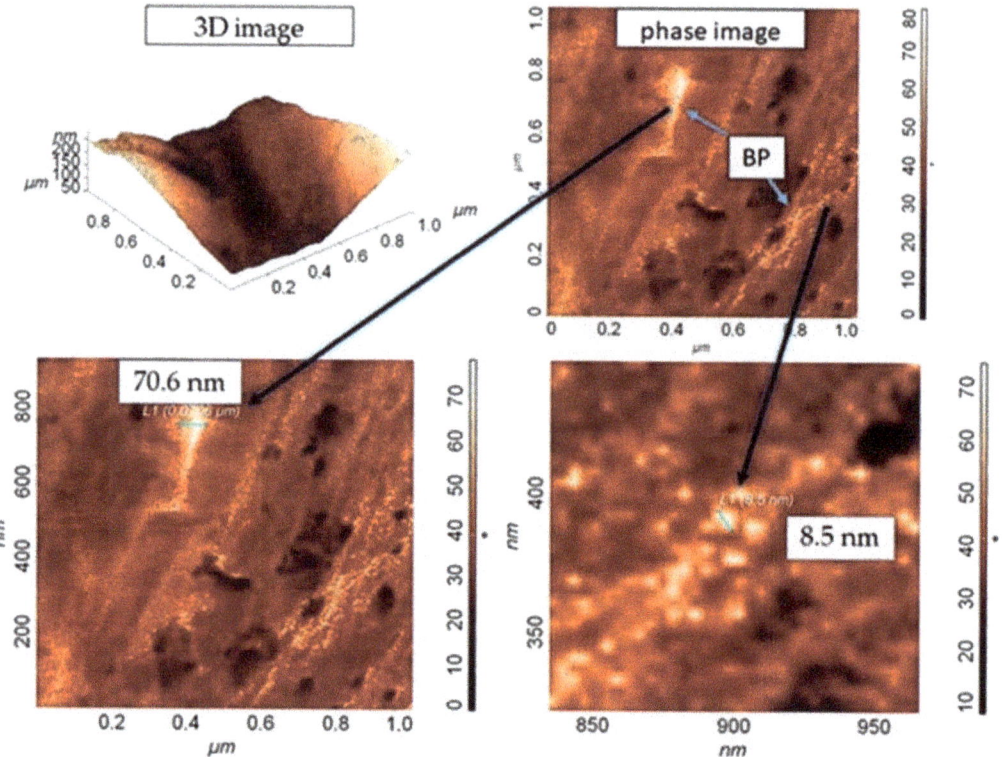

Figure 6. AFM 3D image of topography and phase-contrast of the composite slice, containing 0.5 wt. % BP.

Figure 6 shows 3D images of the topography and phase-contrast of the composite slice containing 0.5 wt. % BP. The choice of this composition of the composite for research on AFM is due to its better mechanical properties. The scanning area was 1 × 1 µm. It was found that the distribution of BP particles in the matrix volume is chaotic. Phase-contrast analysis revealed the presence of a scatter in the sizes of BP particles (from 8.5 nm to ~70 nm). In this case, small BP particles form agglomerates, which, upon crystallization of UHMWPE, orient the crystal growth with the formation of spherulites, where they act as crystallization centers. It was registered that some part of nanosized BP particles are concentrated along the boundaries of spherulite formations due to their migration during pressing. It is known that if a multicomponent material contains several different (non-gaseous) phases, in which at least one of the phases has at least one dimension of the order of nanometers, then it belongs to nanomaterials. Thus, we have shown the formation of a nanocomposite upon the introduction of nanosized BP particles into UHMWPE.

3.2.4. XRC of Composites

Structural studies of UHMWPE and PCM were carried out by X-ray structural analysis (Figure 7). From the X-ray diffraction patterns of all samples, two obvious intense peaks at $2\theta \approx 21.5°$ and $24.0°$ can be distinguished, corresponding to the crystallographic planes (110) and (200) of the UHMWPE polymer [45]; no other peaks were found. The original BP is an amorphous compound as noted above. When BP was injected into UHMWPE, no additional peaks were recorded on PCM radiographs.

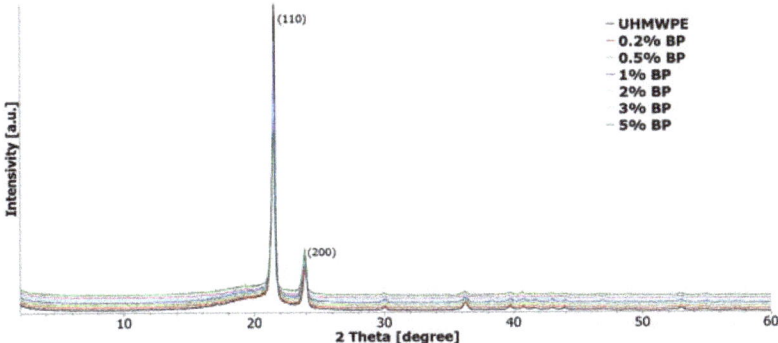

Figure 7. X-ray diffraction patterns of UHMWPE and PCM.

Table 1 shows the results of XRD analysis of UHMWPE and UHMWPE/BP composites.

Table 1. Results of X-ray structural analysis.

Samples	X-ray Structural Analysis			
	α, %	2θ (°)	L, nm	d, nm
initial UHMWPE	58	21.5096	34.15	0.41
UHMWPE + 0.2% BP	56	21.4988	33.35	0.41
UHMWPE + 0.5% BP	55	21.4938	33.17	0.41
UHMWPE + 1% BP	56	21.4933	34.41	0.41
UHMWPE + 2% BP	56	21.4640	32.39	0.41
UHMWPE + 3% BP	56	21.4394	31.75	0.41
UHMWPE + 5% BP	50	21.4547	31.83	0.41

Notes: α—degree of crystallinity, %; 2θ—angle θ, (°); L—crystallite size, nm; d—interplanar distance, nm.

As Table 1 suggests, the introduction of a borpolymer into UHMWPE reduces the degree of crystallinity by 3% at a filler content from 0.2 to 3 wt. %, calculated from the ratio of the intensities of the crystalline and amorphous phases. The degree of crystallinity of the UHMWPE/5 wt. % BP composite decreased by 14% relative to the initial polymer. This may be due to the effect of agglomeration of the filler, which limits the molecular mobility of polymer chains and prevents the crystallization of the polymer [46]. The crystallite sizes of PCM, calculated according to the Scherrer equation at a content of 0.2 to 1 wt. % BP, remain at the level of the initial polymer; with a further increase in the BP content from 2 to 5 wt. % in UHMWPE, a decrease in the crystallite size is observed.

3.2.5. Raman Spectra of Composites

Figure 8 shows the Raman spectra of the initial UHMWPE and the composite containing 5 wt. % BP. Raman spectra are sensitive to vibrations of the crystal lattice (crystalline state) of polyethylene; due to this, these spectra are used to explain the effect of fillers on the phase state of the matrix [47]. In this case, vibrational absorption bands are recorded in the region of 1000 and 1600 cm^{-1}, caused by the twisting of the $\delta(CH_2)$ bond and the stretching of the bonds (CC).

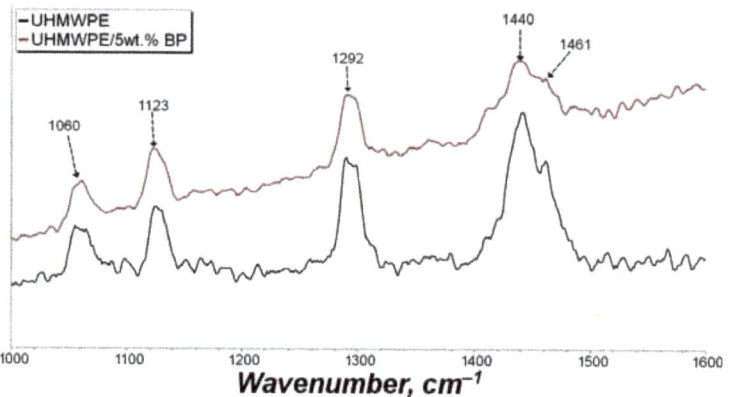

Figure 8. Raman spectra of the initial UHMWPE and composite UHMWPE/5 wt. % BP.

As can be seen from Figure 8, in the Raman spectrum of UHMWPE and PCM, characteristic peaks in the region of 1060 and 1123 cm^{-1} are visible, referring to symmetric and asymmetric stretching vibrations of the C–C bond in the crystalline phase of PE. The peak at 1292 cm^{-1} corresponds to the bending torsional vibrations of the CH_2 group in the crystalline phase. The absorption bands in the region of 1440 and 1461 cm^{-1} refer to bending vibrations of the CH_2 group of the amorphous phase of polyethylene [48,49]. In the Raman spectrum, the indicator of crystallinity of polyethylene is a peak at 1416 cm^{-1}, which is weakly expressed in the initial UHMWPE and PCM. It was found that the introduction of BP into UHMWPE leads to broadening of the absorption band related to vibrations in the amorphous phase. There is also a decrease in the intensity of the 1292 cm^{-1} peak of the τCH_2 crystalline vibration. The results obtained indicate a decrease in the crystallinity of UHMWPE upon the introduction of BP, and, on the whole, agree with the results of X-ray diffraction analysis. Thus, the introduction of BP into UHMWPE leads to a decrease in the crystallinity of the composite.

3.3. Thermodynamic Properties of PCM

Figure 9 shows the DSC data curves obtained by heating the samples, and Table 2 presents the data of the study results.

As evident from Figure 9 and Table 2, the temperature of the onset of melting of PCM does not change over the entire concentration range. Some shift of the melting peaks is observed, but these changes are insignificant and are included in the measurement error range. There is a narrowing of the DSC curves of the composites in comparison with the original UHMWPE.

It was shown that the degree of crystallinity of the initial UHMWPE is 58.7%. After the introduction of BP, a decrease in the degree of crystallinity by 18% is observed. In general, the degree of crystallinity of composites in the entire concentration range of filling is 47–48%. DSC crystallinity values differ from XRC data. However, both methods demonstrate a similar change trend, which is associated with the amorphization of UHMWPE with the introduction of BP, which leads to a decrease in the degree of crystallinity. Thus, the filler affects the growth and shape of crystallites in the process of PCM structuring, which consists in some deformation of the crystalline regions [50].

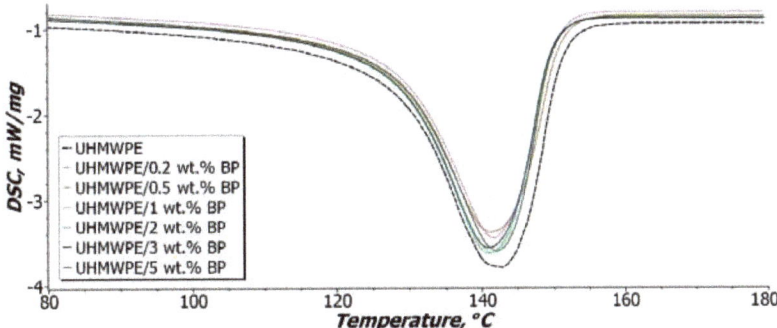

Figure 9. Heating melting function curve of UHMWPE and PCM.

Table 2. Melting point, melting enthalpy, and degree of crystallinity of UHMWPE and composite.

Samples	Thermodynamic Properties		
	T_{onset}, °C	ΔH_{me}, J/g	α, %
initial UHMWPE	127.7	171.1	58.7
UHMWPE + 0.2% BP	128.4	139.3	47.9
UHMWPE + 0.5% BP	128.0	138.5	47.8
UHMWPE + 1% BP	127.6	137.2	47.6
UHMWPE + 2% BP	128.1	138.4	48.5
UHMWPE + 3% BP	127.9	135.7	48.1
UHMWPE + 5% BP	128.4	135.2	48.9

Notes: T_{onset}—melting point onset temperature, °C; ΔH_{me}—melting enthalpy, J/g; α—degree of crystallinity, %.

It was found that the enthalpy of melting of the composites decreases in comparison with the initial UHMWPE. In a series of composites, the enthalpy of melting gradually decreases with an increase in the BP content, which is associated with the loosening of the UHMWPE structure. The authors in [41] argue that the decrease in enthalpy is caused by the nature of the interaction in the compositional system. If the interaction between the polymer and the filler prevails, where the active surface of the filler acts as a nucleating agent (to heterogeneous nucleation) during crystallization, this leads to an increase in the enthalpy of melting and the degree of crystallinity. This tendency is observed in heterogeneous systems, where the filler has a high surface activity [51–54]. In the case of a predominant interaction between filler particles, the formation of agglomerates is observed, which limits the rate of polymer crystallization. Thus, it was shown that with an increase in the BP content in PCM, agglomeration between filler particles intensifies, which leads to a decrease in the enthalpy of melting by 20% compared to the initial UHMWPE. In this case, the formation of less perfect and defective structural elements—spherulites is observed in the supramolecular structure of PCM (Figure 5d,e). Nevertheless, a decrease in the degree of crystallinity and enthalpy of melting does not lead to a deterioration in the mechanical properties of the composite. It is known that the amorphous phase in UHMWPE contributes to an increase in the impact toughness of the material, due to the effect of linkage of through-feed chains [55].

Thus, the introduction of BP contributes to an overall decrease in the degree of crystallinity and the enthalpy of fusion of UHMWPE.

3.4. Mechanical Properties of PCM

BP is actively used as a hardening modifier for thermosetting plastics and industrial rubber goods, which explains the increased interest in this material. The mechanical characteristics of UHMWPE filled by BP are presented in Table 3 and Figure 10.

Table 3. Elongation at break, tensile strength, and Young's modulus of UHMWPE and PCM with borpolymer (BP).

Samples	σ_T, MPa	ε_b, %	E, MPa
initial UHMWPE	32 ± 3	339 ± 16	420 ± 26
UHMWPE + 0.2% BP	49 ± 1	434 ± 14	472 ± 34
UHMWPE + 0.5% BP	50 ± 1	417 ± 10	524 ± 37
UHMWPE + 1% BP	45 ± 1	389 ± 8	499 ± 19
UHMWPE + 2% BP	43 ± 2	383 ± 18	519 ± 32
UHMWPE + 3% BP	43 ± 1	369 ± 12	524 ± 21
UHMWPE + 5% BP	39 ± 1	327 ± 14	520 ± 22

Notes: σ_T—tensile strength, MPa; ε_b—elongation at break, %; E—Young's modulus in deformation 0.1–0.3%, MPa.

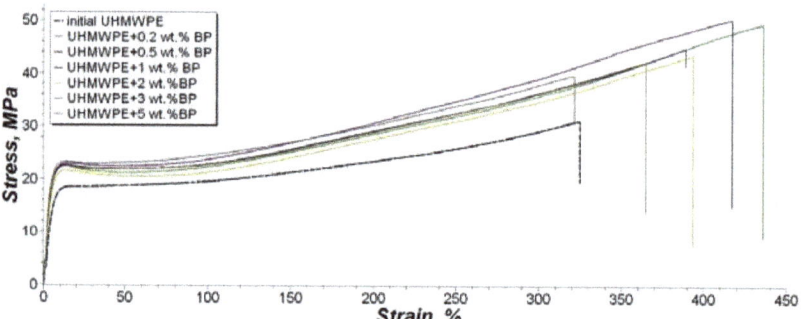

Figure 10. Stress–strain curve of the tensile tests.

For the stress–strain curve, we took the test data of composites, which are similar to the average values after statistic processing.

Analysis of the results of PCM mechanical characteristics showed that 0.2 and 0.5 wt. % BP content leads to a significant increase in the strength and elasticity of the material. The increase in tensile strength of PCM was noted by 53% and 56% relative to the original polymer, respectively. At the same time, there is an increase in the elongation at break by 28% and 23%. A further increase in BP content leads to a gradual decrease in these parameters. However, the value of the elongation at break of the composite containing 5 wt. % BP, remains within the measurement error. The tensile strength of the UHMWPE/5 wt. % BP composite is 18% higher compared to unfilled UHMWPE. The modulus of elasticity of the composites and the original UHMWPE does not undergo significant changes, which indicates that the rigidity of the material is preserved throughout the entire concentration range of filling.

Based on studies of the supramolecular structure of PCM, it was found that the introduction of low concentrations of BP forms a fine-spherulite structure, which explains the maximum increase in mechanical parameters. At high concentrations, the occurrence of defective areas is observed, which leads to a slight decrease in mechanical parameters relative to PCM with a lower BP content, and does not decrease as compared to the original UHMWPE.

Studies on the modification of the UHMWPE matrix with thermosetting polymers or organic compounds of the ester class are poorly understood. In addition, the use of this borpolymer as a thermoplastic modifier has not been previously considered. Wang et al. showed the effect of thermosetting polymers on the wettability of UHMWPE fibers [56,57]. It was found that thermosetting binders of various types increase the wettability of UHMWPE fibers, thereby increasing the strength of the material by enhancing the adhesive interaction between the components of the composite. In [11] it was found that polyphenyl ether combined with carbon fibers increases the wear resistance of PCM, also due to the enhancement of interfacial interaction between the components, and due to the participation of ether in the formation of secondary structures on the friction surface. BP is known to be used effectively in elastomeric materials as a reinforcing agent. In this case, BP acts as a modifier of the rubber matrix, contributing to the formation of a stable three-dimensional vulcanization network. Moreover, the presence of a boron atom enhances the interaction at the polymer–filler interface, which indicates the reactivity of BP during vulcanization [35]. However, the results of IR spectroscopy of the UHMWPE/5 wt. % BP composite (Figure 4) indicate that the filler particles do not interact with the polymer macromolecule. Thus, BP acts as a reinforcing modifier for the polymer matrix.

In addition to the effect of strengthening the polymer matrix, an increase in the deformation and strength characteristics is due to the formation of the spherulite structure of PCM [4,58]. It is known [59] that PCMs with small spherulites are usually more rigid than composites consisting of large spherulites (Figure 5). Mechanical deformation of composites with a spherulite structure first destroys the boundary regions of the spherulites, i.e., the interlamellar amorphous part. Then the inner part of the spherulites undergoes deformation, since the crystalline ordered phase of the polymer is stronger [60]. Thus, composites characterized by a large amount of spherulites, for example, in a fine-spherulite structure, will have increased strength, while in composites characterized by the formation of an inhomogeneous and coarse-spherulite structure, the boundary regions are usually weak [58]; therefore, with an increase in the filler content, the occurrence of defective regions in the supramolecular structure of PCM is observed, which is accompanied by a slight decrease in mechanical parameters (Figure 5d,e). In [61], data are provided showing that organic fillers with a low molecular weight plasticize the UHMWPE matrix during stretching, facilitating relaxation processes. The increase in the relative elongation of composites containing BP can be explained by the plasticizing effect of BP.

The results of studying the effect of borpolymer on the compressive strength of PCM at different relative deformations are presented in Table 4 and Figure 11.

Table 4. Melting point, melting enthalpy, and degree of crystallinity of UHMWPE and composite.

Samples	Compressive Strength		
	2.5% Deformation	10% Deformation	25% Deformation
initial UHMWPE	4 ± 1	17 ± 2	30 ± 1
UHMWPE + 0.2% BP	11 ± 2	28 ± 2	29 ± 2
UHMWPE + 0.5% BP	9 ± 1	21 ± 1	26 ± 2
UHMWPE + 1% BP	9 ± 2	23 ± 1	30 ± 1
UHMWPE + 2% BP	13 ± 2	26 ± 2	31 ± 1
UHMWPE + 3% BP	11 ± 1	24 ± 3	33 ± 1
UHMWPE + 5% BP	10 ± 2	25 ± 1	34 ± 1

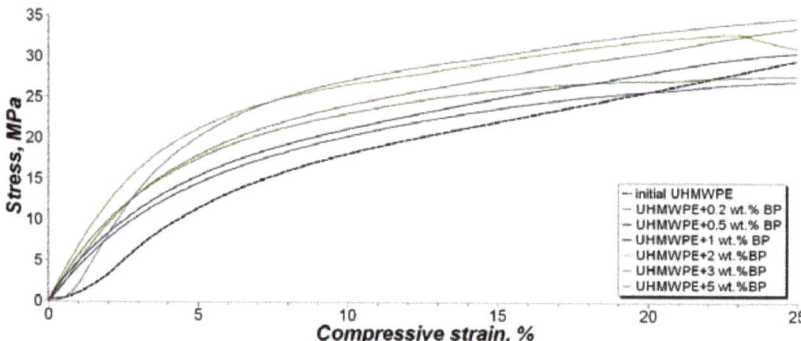

Figure 11. Stress–strain curve of the compressive tests.

It was found that the introduction of BP into the polymer leads to an increase in compressive stress, at a specified relative deformation of 2.5%, by about 2–3 times compared with the initial UHMWPE. High values of the compressive stress at a specified relative deformation of 10% are observed for the composite with the composition UHMWPE/0.2 wt. % BP and UHMWPE/2 wt. % BP, in which an increase of 65% and 53% is noted, respectively. The compressive stress, at compressive strength at 25% strain of the composites, changes insignificantly depending on the filler content and remains within the measurement error. The increase in compressive strength values is attributed to an increase in the material's resistance to deformation during compression, due to the formation of a reinforced PCM system [62]. In addition, it is assumed that, due to the high molecular weight of UHMWPE, regions with large overlaps of long macromolecule chains are formed. The occurrence of such zones with large overlap increases the ability of PCM to transfer a large compressive force from molecule to molecule [63].

Thus, the introduction of BP into the polymer leads to an increase in the deformation-strength characteristics and compressive strength, even at a low filler content.

4. Conclusions

The effect of borpolymer on the mechanical properties and structure of UHMWPE has been studied. It was found that the use of borpolymer as a UHMWPE modifier made it possible to increase the mechanical characteristics of the material at low BP concentrations (at 0.2 and 0.5 wt. %). At these concentrations, a maximum increase in tensile strength of 56% and elongation at break of 28%, relative to the original UHMWPE, was recorded. An increase in compressive strength was established at a specified relative deformation of 2.5% and 10% in the entire concentration range of PCM; the maximum values of these indicators were 13 MPa and 28 MPa, respectively. No significant changes in the modulus of elasticity are observed. The study of the processes of structure formation by the SEM method revealed the formation of a spherulite structure upon the introduction of BP, which explains the increase in the tensile strength of PCM. By means of IR spectroscopy, it was found that the borpolymer does not enter into chemical interactions with UHMWPE during processing. The presence of the main peaks of absorption caused by the vibrations of bonds of the initial components UHMWPE and BP was found. DSC and XRC studies of the degree of crystallinity revealed a general decrease in this parameter caused by loosening and amorphization of the structure with increasing BP concentration. These changes lead to a decrease in the enthalpy of melting by 20% compared to the initial polymer. The increase in the elasticity of the material is explained by the fact that the introduction of an amorphous filler into UHMWPE facilitates relaxation processes when an external load is applied.

Thus, BP is an effective filler for UHMWPE, helping to increase the tensile strength and elongation of a composite.

Author Contributions: Conceptualization, A.A.D.; methodology, S.N.D., A.P.V. and A.A.D.; software, S.N.D. and A.A.K.; validation, A.A.O. and A.G.T.; formal analysis, A.P.V., A.A.O. and S.N.D.; investigation, A.P.V. and A.A.D.; resources, A.A.O. and A.K.K.; data curation, A.A.D.; writing—original draft preparation, S.N.D.; writing—review & editing, A.P.V. and A.A.O.; visualization, S.N.D.; supervision, A.K.K. and A.G.T.; project administration, A.A.D.; funding acquisition, A.A.O. All authors have read and agreed to the published version of the manuscript.

Funding: The paper was prepared as part of the implementation of the state assignment for scientific research by laboratories under the guidance of young, promising researchers in the framework of the national project "Science and Universities" No. 121112400007-5.

Data Availability Statement: The data used to support the findings of this study are available from the corresponding author upon request.

Conflicts of Interest: The authors declare no conflict of interest. The funding bodies had no role in the design of the study, the collection, analysis, or interpretation of data, in the writing of the manuscript, or in the decision to publish the results.

References

1. Kurtz, S.M. *The UHMWPE Handbook: Ultra-High Molecular Weight Polyethylene in Total Joint Replacement*; Springer: New York, NY, USA, 2009.
2. Hussain, M.; Naqvi, R.A.; Abbas, N.; Khan, S.M.; Nawaz, S.; Hussain, A.; Zahra, N.; Khalid, M.W. Ultra-High-Molecular-Weight-Polyethylene (UHMWPE) as a Promising Polymer Material for Biomedical Applications: A Concise Review. *Polymers* **2020**, *12*, 323. [CrossRef] [PubMed]
3. Cummings, C.S.; Lucas, E.M.; Marro, J.A.; Kieu, T.M.; DesJardins, J.D. The Effects of Proton Radiation on UHMWPE Material Properties for Space Flight and Medical Applications. *Adv. Space Res.* **2011**, *48*, 1572–1577. [CrossRef]
4. Danilova, S.N.; Yarusova, S.B.; Kulchin, Y.N.; Zhevtun, I.G.; Buravlev, I.Y.; Okhlopkova, A.A.; Gordienko, P.S.; Subbotin, E.P. UHMWPE/CaSiO3 Nanocomposite: Mechanical and Tribological Properties. *Polymers* **2021**, *13*, 570. [CrossRef]
5. Panin, S.V.; Buslovich, D.G.; Dontsov, Y.V.; Bochkareva, S.A.; Kornienko, L.A.; Berto, F. UHMWPE-Based Glass-Fiber Composites Fabricated by FDM. Multiscaling Aspects of Design, Manufacturing and Performance. *Materials* **2021**, *14*, 1515. [CrossRef] [PubMed]
6. Panin, S.V.; Kornienko, L.A.; Valentyukevich, N.N.; Alexenko, V.O.; Ovechkin, B.B. Mechanical and Tribotechnical Properties of Three-Component Solid Lubricant UHMWPE Composites. In *AIP Conference Proceedings*; AIP Publishing LLC: Ekaterinburg, Russia, 2018; p. 030050. [CrossRef]
7. Cao, Z.; Shi, G.; Yan, X.; Wang, Q. In Situ Fabrication of CuO/UHMWPE Nanocomposites and Their Tribological Performance. *J. Appl. Polym. Sci.* **2019**, *136*, 47925. [CrossRef]
8. Cao, S.; Liu, H.; Ge, S.; Wu, G. Mechanical and Tribological Behaviors of UHMWPE Composites Filled with Basalt Fibers. *J. Reinf. Plast. Compos.* **2011**, *30*, 347–355. [CrossRef]
9. Dangsheng, X. Friction and Wear Properties of UHMWPE Composites Reinforced with Carbon Fiber. *Mater. Lett.* **2005**, *59*, 175–179. [CrossRef]
10. Panin, S.V.; Kornienko, L.A.; Huang, Q.; Buslovich, D.G.; Bochkareva, S.A.; Alexenko, V.O.; Panov, I.L.; Berto, F. Effect of Adhesion on Mechanical and Tribological Properties of Glass Fiber Composites, Based on Ultra-High Molecular Weight Polyethylene Powders with Various Initial Particle Sizes. *Materials* **2020**, *13*, 1602. [CrossRef]
11. Wang, Z.; Ma, Y.; Guo, L.; Tong, J. Influence of Polyphenyl Ester and Nanosized Copper Filler on the Tribological Properties of Carbon Fibre–Reinforced Ultra-High-Molecular-Weight Polyethylene Composites. *J. Thermoplast. Compos. Mater.* **2018**, *31*, 1483–1496. [CrossRef]
12. Panin, S.V.; Alexenko, V.O.; Buslovich, D.G.; Anh, N.D.; Qitao, H. Solid-Lubricant, Polymer—Polymeric and Functionalized Fiber—And Powder Reinforced Composites of Ultra-High Molecular Weight Polyethylene. *IOP Conf. Ser. Earth Environ. Sci.* **2018**, *115*, 012010. [CrossRef]
13. Cheng, B.; Duan, H.; Chen, S.; Shang, H.; Li, J.; Shao, T. Phase Morphology and Tribological Properties of PI/UHMWPE Blend Composites. *Polymer* **2020**, *202*, 122658. [CrossRef]
14. Panin, S.V.; Kornienko, L.A.; Nguen Suan, T.; Ivanova, L.R.; Korchagin, M.A.; Shil'ko, S.V.; Pleskachevskii, Y.M. Wear Resistance of Composites Based on Hybrid UHMWPE–PTFE Matrix: Mechanical and Tribotechnical Properties of the Matrix. *J. Frict. Wear* **2015**, *36*, 249–256. [CrossRef]
15. Silverstein, M.S.; Breitner, J. A Polytetrafluoroethylene Filled Ultra-High Molecular Weight Polyethylene Composite: Mechanical and Wear Property Relationships. *Polym. Eng. Sci.* **1995**, *35*, 1785–1794. [CrossRef]
16. Liu, Y.; Sinha, S.K. Mechanical and Tribological Properties of PEEK Particle-Filled UHMWPE Composites: The Role of Counterface Morphology Change in Dry Sliding Wear. *J. Reinf. Plast. Compos.* **2013**, *32*, 1614–1623. [CrossRef]
17. Cowie, R.M.; Briscoe, A.; Fisher, J.; Jennings, L.M. Wear and Friction of UHMWPE-on-PEEK OPTIMA™. *J. Mech. Behav. Biomed. Mater.* **2019**, *89*, 65–71. [CrossRef]

18. Senatov, F.S.; Chubrik, A.V.; Maksimkin, A.V.; Kolesnikov, E.A.; Salimon, A.I. Comparative Analysis of Structure and Mechanical Properties of Porous PEEK and UHMWPE Biomimetic Scaffolds. *Mater. Lett.* **2019**, *239*, 63–66. [CrossRef]
19. Cowie, R.M.; Pallem, N.M.; Briscoe, A.; Jennings, L.M. Third Body Wear of UHMWPE-on-PEEK-OPTIMA™. *Materials* **2020**, *13*, 1264. [CrossRef]
20. Mohammed, A.S.; Fareed, M.I. Surface Modification of Polyether Ether Ketone (PEEK) with a Thin Coating of UHMWPE for Better Tribological Properties. *Tribol. Trans.* **2017**, *60*, 881–887. [CrossRef]
21. Okhlopkova, T.A.; Borisova, R.V.; Nikiforov, L.A.; Spiridonov, A.M.; Okhlopkova, A.A.; Jeong, D.-Y.; Cho, J.-H. Supramolecular Structure and Mechanical Characteristics of Ultrahigh-Molecular-Weight Polyethylene-Inorganic Nanoparticle Nanocomposites: UHMWPE-Inorganic Nanoparticle Composites. *Bull. Korean Chem. Soc.* **2016**, *37*, 439–444. [CrossRef]
22. Yin, X.; Li, S.; He, G.; Feng, Y.; Wen, J. Preparation and Characterization of CNTs/UHMWPE Nanocomposites via a Novel Mixer under Synergy of Ultrasonic Wave and Extensional Deformation. *Ultrason. Sonochem.* **2018**, *43*, 15–22. [CrossRef]
23. Zhang, H.; Wang, L.; Chen, Q.; Li, P.; Zhou, A.; Cao, X.; Hu, Q. Preparation, Mechanical and Anti-Friction Performance of MXene/Polymer Composites. *Mater. Des.* **2016**, *92*, 682–689. [CrossRef]
24. Zec, J.; Tomić, N.Z.; Zrilić, M.; Lević, S.; Marinković, A.; Heinemann, R.J. Optimization of Al_2O_3 Particle Modification and UHMWPE Fiber Oxidation of EVA Based Hybrid Composites: Compatibility, Morphological and Mechanical Properties. *Compos. Part B Eng.* **2018**, *153*, 36–48. [CrossRef]
25. Wang, Y.; Qiao, X.; Wan, J.; Xiao, Y.; Fan, X. Preparation of AlN Microspheres/UHMWPE Composites for Insulating Thermal Conductors. *RSC Adv.* **2016**, *6*, 80262–80267. [CrossRef]
26. Feng, C.P.; Chen, L.; Wei, F.; Ni, H.Y.; Chen, J.; Yang, W. Highly Thermally Conductive UHMWPE/Graphite Composites with Segregated Structures. *RSC Adv.* **2016**, *6*, 65709–65713. [CrossRef]
27. Silva, C.; Lago, R.; Veloso, H.; Patricio, P. Use of Amphiphilic Composites Based on Clay/Carbon Nanofibers as Fillers in UHMWPE. *J. Braz. Chem. Soc.* **2017**, *29*, 278–284. [CrossRef]
28. Chen, R.; Ye, C.; Xin, Z.; Zhao, S.; Xia, J.; Meng, X. The Effects of Octadecylamine Functionalized Multi-Wall Carbon Nanotubes on the Conductive and Mechanical Properties of Ultra-High Molecular Weight Polyethylene. *J. Polym. Res.* **2018**, *25*, 135. [CrossRef]
29. Yeh, J.; Wang, C.K.; Tsai, C.-C.; Ren, D.-H.; Chiu, S.-H. Ultradrawing and Ultimate Tensile Properties of Novel Ultra-High Molecular Weight Polyethylene Composite Fibers Filled with Nanoalumina Fillers. *Text. Res. J.* **2016**, *86*, 1768–1787. [CrossRef]
30. Abdalla, M.O.; Ludwick, A.; Mitchell, T. Boron-Modified Phenolic Resins for High Performance Applications. *Polymer* **2003**, *44*, 7353–7359. [CrossRef]
31. Qiu, F.; Zhao, W.; Han, S.; Zhuang, X.; Lin, H.; Zhang, F. Recent Advances in Boron-Containing Conjugated Porous Polymers. *Polymers* **2016**, *8*, 191. [CrossRef]
32. Kurt, R.; Mengeloglu, F.; Meric, H. The Effects of Boron Compounds Synergists with Ammonium Polyphosphate on Mechanical Properties and Burning Rates of Wood-HDPE Polymer Composites. *Eur. J. Wood Prod.* **2012**, *70*, 177–182. [CrossRef]
33. Yi, X.; Feng, A.; Shao, W.; Xiao, Z. Synthesis and Properties of Graphene Oxide–Boron-Modified Phenolic Resin Composites. *High Perform. Polym.* **2016**, *28*, 505–517. [CrossRef]
34. Wang, D.-C.; Chang, G.-W.; Chen, Y. Preparation and Thermal Stability of Boron-Containing Phenolic Resin/Clay Nanocomposites. *Polym. Degrad. Stab.* **2008**, *93*, 125–133. [CrossRef]
35. Korabel'nikov, D.V.; Lenskii, M.A.; Ozhogin, A.V. Study of the Modifying Effect of Additions of Boric Acid Polymethylene-P-Triphenyl Ester in Rubber-Based Polymer Composites. *Int. Polym. Sci. Technol.* **2012**, *39*, 17–20. [CrossRef]
36. Korabel'nikov, D.V.; Lenskii, M.A.; Ozhogin, A.V.; Nartov, A.S.; Anan'eva, E.S. A Study of the Modifying Effect of Additions of Boric Acid Polymethylene- p -Triphenyl Ester in Rubber-Based Polymer Composites. Part 3. *Int. Polym. Sci. Technol.* **2016**, *43*, 11–14. [CrossRef]
37. Androshchuk, A.A.; Lenskii, M.A.; Belousov, A.M. The Interaction of Polyesters and Polymethylene Esters of Phenols and Boric Acid with Epoxy Resin. *Int. Polym. Sci. Technol.* **2011**, *38*, 33–36. [CrossRef]
38. Korabel'nikov, D.V.; Lenskii, M.A.; Nekrasov, M.S.; Kondrat'ev, R.N.; Kartavykh, I.E. Increasing the Strength and Wear Resistance of Friction Composite Materials by Modifying Them with Boric Acid Polymethylene-P-Triphenyl Ester. *Int. Polym. Sci. Technol.* **2013**, *40*, 51–55. [CrossRef]
39. Zimin, D.E. Thermal and Chemical Stability of Glass Fibers with a Boron Polymer Protective Coating. *Glass Ceram.* **2015**, *72*, 51–53. [CrossRef]
40. Lenskiy, M.A.; Shul'ts, E.E.; Korabel'nikov, D.V.; Ozhogin, A.V.; Novitskiy, A.N. Synthesis of Polyesters of Diatomic Phenols and Boric Acid and Their Interaction with Formaldehyde. *Polym. Sci. Ser. B* **2019**, *61*, 530–539. [CrossRef]
41. Boon Peng, C.; Hazizan, M.A.; Nasir, R.M. The Effect of Zeolite on the Crystallization Behaviour and Tribological Properties of UHMWPE Composite. *AMR* **2013**, *812*, 100–106. [CrossRef]
42. Nandiyanto, A.B.D.; Oktiani, R.; Ragadhita, R. How to Read and Interpret FTIR Spectroscope of Organic Material. *Indones. J. Sci. Technol.* **2019**, *4*, 97–118. [CrossRef]
43. Shurvell, H.F.; Faniran, J.A. Infrared Spectra of Triphenylboron and Triphenylborate. *Can. J. Chem.* **1968**, *46*, 2081–2087. [CrossRef]
44. Garcia, R.; Proksch, R. Nanomechanical Mapping of Soft Matter by Bimodal Force Microscopy. *Eur. Polym. J.* **2013**, *49*, 1897–1906. [CrossRef]
45. Joo, Y. Characterization of Ultra High Molecular Weight Polyethyelene Nascent Reactor Powders by X-ray Diffraction and Solid State NMR. *Polymer* **2000**, *41*, 1355–1368. [CrossRef]

46. De Oliveira Aguiar, V.; Pita, V.J.R.R.; De Fatima Vieira Marques, M.; Soares, I.T.; Martins Ferreira, E.H.; Oliveira, M.S.; Monteiro, S.N. Ultra-High Molecular Weight Polyethylene Nanocomposites Reinforced with Novel Surface Chemically Modified Sonic-Exfoliated Graphene. *J. Mater. Res. Technol.* **2021**, *11*, 1932–1941. [CrossRef]
47. Rull, F.; Prieto, A.C.; Casado, J.M.; Sobron, F.; Edwards, H.G.M. Estimation of Crystallinity in Polyethylene by Raman Spectroscopy. *J. Raman Spectrosc.* **1993**, *24*, 545–550. [CrossRef]
48. Kotula, A.P.; Meyer, M.W.; DeVito, F.; Plog, J.; Hight Walker, A.R.; Migler, K.B. The Rheo-Raman Microscope: Simultaneous Chemical, Conformational, Mechanical, and Microstructural Measures of Soft Materials. *Rev. Sci. Instrum.* **2016**, *87*, 105105. [CrossRef]
49. Affatato, S.; Modena, E.; Carmignato, S.; Taddei, P. The Use of Raman Spectroscopy in the Analysis of UHMWPE Uni-Condylar Bearing Systems after Run on a Force and Displacement Control Knee Simulators. *Wear* **2013**, *297*, 781–790. [CrossRef]
50. Sobieraj, M.C.; Rimnac, C.M. Ultra High Molecular Weight Polyethylene: Mechanics, Morphology, and Clinical Behavior. *J. Mech. Behav. Biomed. Mater.* **2009**, *2*, 433–443. [CrossRef] [PubMed]
51. Chang, B.P.; Akil, H.M.; Nasir, R.B.; Khan, A. Optimization on Wear Performance of UHMWPE Composites Using Response Surface Methodology. *Tribol. Int.* **2015**, *88*, 252–262. [CrossRef]
52. Wu, Z.; Zhang, Z.; Mai, K. Non-Isothermal Crystallization Kinetics of UHMWPE Composites Filled by Oligomer-Modified $CaCO_3$. *J. Therm. Anal. Calorim.* **2020**, *139*, 1111–1120. [CrossRef]
53. Efe, G.C.; Bindal, C.; Ucisik, A.H. Characterization of UHWPE-TiO_2 Composites Produced by Gelation/Crystallization Method. *Acta Phys. Pol. A* **2017**, *132*, 767–769. [CrossRef]
54. Jafari, I.; Shakiba, M.; Khosravi, F.; Ramakrishna, S.; Abasi, E.; Teo, Y.S.; Kalaee, M.; Abdouss, M.; Ramazani, S.A.A.; Moradi, O.; et al. Thermal Degradation Kinetics and Modeling Study of Ultra High Molecular Weight Polyethylene (UHMWP)/Graphene Nanocomposite. *Molecules* **2021**, *26*, 1597. [CrossRef] [PubMed]
55. Peacock, A.J. *Handbook of Polyethylene: Structures, Properties and Applications*; Marcel Dekker, Inc.: New York, NY, USA, 2000.
56. Zhu, D.; Wang, Y.; Zhang, X.; Cheng, S. Interfacial Bond Property of UHMWPE Composite. *Polym. Bull.* **2010**, *65*, 35–44. [CrossRef]
57. Zhang, X.; Wang, Y.; Cheng, S. Properties of UHMWPE Fiber-Reinforced Composites. *Polym. Bull.* **2013**, *70*, 821–835. [CrossRef]
58. Way, J.L.; Atkinson, J.R.; Nutting, J. The Effect of Spherulite Size on the Fracture Morphology of Polypropylene. *J. Mater. Sci.* **1974**, *9*, 293–299. [CrossRef]
59. Khalil, Y.; Hopkinson, N.; Kowalski, A.; Fairclough, J.P.A. Characterisation of UHMWPE Polymer Powder for Laser Sintering. *Materials* **2019**, *12*, 3496. [CrossRef]
60. Butler, M.F.; Donald, A.M. Deformation of spherulitic polyethylene thin films. *J. Mater. Sci.* **1997**, *32*, 3675–3685. [CrossRef]
61. Dayyoub, T.; Olifirov, L.K.; Chukov, D.I.; Kaloshkin, S.D.; Kolesnikov, E.; Nematulloev, S. The Structural and Mechanical Properties of the UHMWPE Films Mixed with the PE-Wax. *Materials* **2020**, *13*, 3422. [CrossRef]
62. Petrova, P.N.; Gogoleva, O.V.; Argunova, A.G. Development of Polymer Composites Based on Ultrahigh Molecular Weight Polyethylene, Polytetrafluoroethylene and Carbon Fibers. In *AIP Conference Proceedings*; AIP Publishing LLC: Ekaterinburg, Russia, 2018; p. 030052. [CrossRef]
63. Tulatorn, V.; Ouajai, S.; Yeetsorn, R.; Chanunpanich, N. Mechanical Behavior Investigation of UHMWPE Composites for Pile Cushion Applications. *KMUTNB IJAST* **2015**, *8*, 1–12. [CrossRef]

Article

CuO/PMMA Polymer Nanocomposites as Novel Resist Materials for E-Beam Lithography

Georgia Geka [1,2], George Papageorgiou [2], Margarita Chatzichristidi [1,*], Andreas Germanos Karydas [3], Vassilis Psycharis [2] and Eleni Makarona [2,*]

1. Department of Chemistry, National and Kapodistrian University of Athens, Zografou, 157 71 Athens, Greece; georgia.geka101@gmail.com
2. Institute of Nanoscience and Nanotechnology, NCSR "Demokritos", Aghia Paraskevi, 153 10 Athens, Greece; g.papageorgiou@inn.demokritos.gr (G.P.); v.psycharis@inn.demokritos.gr (V.P.)
3. Institute of Nuclear and Particle Physics, NCSR "Demokritos", 153 10 Athens, Greece; karydas@inp.demokritos.gr
* Correspondence: mchatzi@chem.uoa.gr (M.C.); e.makarona@inn.demokritos.gr (E.M.); Tel.: +30-210-7274335 (M.C.); +30-210-6503662 (E.M.)

Citation: Geka, G.; Papageorgiou, G.; Chatzichristidi, M.; Karydas, A.G.; Psycharis, V.; Makarona, E. CuO/PMMA Polymer Nanocomposites as Novel Resist Materials for E-Beam Lithography. *Nanomaterials* **2021**, *11*, 762. https://doi.org/10.3390/nano11030762

Academic Editor: Teresa Cuberes

Received: 12 February 2021
Accepted: 16 March 2021
Published: 17 March 2021

Publisher's Note: MDPI stays neutral with regard to jurisdictional claims in published maps and institutional affiliations.

Copyright: © 2021 by the authors. Licensee MDPI, Basel, Switzerland. This article is an open access article distributed under the terms and conditions of the Creative Commons Attribution (CC BY) license (https://creativecommons.org/licenses/by/4.0/).

Abstract: Polymer nanocomposites have emerged as a new powerful class of materials because of their versatility, adaptability and wide applicability to a variety of fields. In this work, a facile and cost-effective method to develop poly(methyl methacrylate) (PMMA)-based polymer nanocomposites with copper oxide (CuO) nanofillers is presented. The study concentrates on finding an appropriate methodology to realize CuO/PMMA nanocomposites that could be used as resist materials for e-beam lithography (EBL) with the intention of being integrated into nanodevices. The CuO nanofillers were synthesized via a low-cost chemical synthesis, while several loadings, spin coating conditions and two solvents (acetone and methyl ethyl ketone) were explored and assessed with regards to their effect on producing CuO/PMMA nanocomposites. The nanocomposite films were patterned with EBL and contrast curve data and resolution analysis were used to evaluate their performance and suitability as a resist material. Micro-X-ray fluorescence spectroscopy (μ-XRF) complemented with XRF measurements via a handheld instrument (hh-XRF) was additionally employed as an alternative rapid and non-destructive technique in order to investigate the uniform dispersion of the nanofillers within the polymer matrix and to assist in the selection of the optimum preparation conditions. This study revealed that it is possible to produce low-cost CuO/PMMA nanocomposites as a novel resist material without resorting to complicated preparation techniques.

Keywords: polymer nanocomposites; CuO nanostructures; PMMA; e-beam lithography; resist process engineering; X-ray fluorescence; chemical synthesis

1. Introduction

Polymer nanocomposites, defined as polymers containing fillers with at least one dimension smaller than 100 nm at very low loadings (<5 vol. %), have emerged as a new and promising class of materials for a wide range of applications, which may span from the automotive [1] to the textile industry [2] or even bioengineering [3]. More importantly, polymer nanocomposites can be used as alternative building blocks or as the functional core of novel micro/nano-electronic devices [4–8]. This broadening of applicability is due to their very nature. In contrast to traditional polymer composites with high loadings (>50% *w/v*) of micrometer-size fillers, which have been used for almost 100 years [9], the recent progress of nanotechnology has provided a plethora of nanofillers, which, even at low % vol. loadings can drastically enhance and modify the polymer's properties with respect to its bulk counterpart [10]. Hence, research efforts into polymer nanocomposites has revolved around the successful incorporation of nano-sized fillers into polymers so as to take full advantage of the nanofillers' multi-faceted nature and to develop a new class of organic/inorganic materials of enhanced properties and multi-functionality [11,12]. Of particular interest is the fact that the newly-developed properties of the nanocomposites are largely different due to the actual morphology of the selected nanofillers and strongly depend on whether the nanofillers are two-dimensional (2D) layered structures, one-dimensional (1D) fibrous or zero-dimensional (0D) spherical ones [10]. Another critical parameter when creating polymer nanocomposites is the uniform dispersion of these isotropic or anisotropic nano-sized fillers, since the distribution itself controls the ultra-large interfacial area per unit volume between nano-scale fillers and host polymers, which in itself dictates the composite's properties [10].

However, a major hurdle in the development of polymer nanocomposites, and in particular of nanocomposites of inorganic fillers, such as metal oxide nanoparticles, is the difficulty in obtaining homogeneous dispersions within the polymer matrix and in preventing agglomeration/aggregation and sedimentation of the nanofillers [13–15]. This difficulty mainly arises from two facts: (1) nanofillers are typically hydrophilic, while polymers are typically hydrophobic, and (2) nanofillers being very small in size (<100 nm) tend to agglomerate in order to minimize the surface to volume ratio, and in turn, the surface free energy of the system. The driving force for the agglomeration process is the van der Waals attraction. In general, the agglomerates are hard to break and do not produce the intended properties' enhancement, even when dispersed in the polymer matrix in a homogeneous manner.

This work has concentrated on the development of a relatively facile approach for the creation of CuO/Poly(methyl methacrylate) (PMMA) nanocomposites as a novel resist material for electron beam lithography (EBL) with the final aim of being employed as the functional material in nanoelectronic devices, such as gas sensors. In most reported works, such as the those of the Gonsalves group since the early 2000s [16–18], the incorporation of nanofillers of sizes less than 10 nm inside resist hosts is performed with the aim of improving the resolution without sacrificing the inherent sensitivity and contrast. However, the incentive of the present work is to add functionalities to the resist by the inclusion of CuO nanostructures with typical sizes of 10–30 nm and to explore whether the composite PMMA can still be patterned by EBL and to what extent its properties (resolution, sensitivity) may be affected.

While grafting has been the most common method to enhance the nanofiller miscibility (e.g., References [19,20]), the suggested approach in this work entails the use of polar solvents and physical mixing so as to keep the cost and preparation time as low as possible, and it was based on the reported results of Botsi et al. [21]. Physical mixing of CuO nanofillers with the polymer host has also been reported in other works [22–24], but in all cases, the films were merely formed by drop-casting and lithographic patterning was not attempted. CuO was selected as the nanofiller as it has been proven to be one of the most promising and versatile metal oxides exhibiting a remarkable spectrum of properties, such as catalytic activity [25], energy storage capabilities [26], optoelectronic [27,28] and antibacterial properties [29], and most notably gas sensing properties [30–32]. This work was mainly devised with the latter in mind so as to create a novel CuO/PMMA resist material readily applicable for the development of polymer-based novel gas sensors [24,33]. PMMA was chosen as the polymer host, since it is one of the most widely employed resists in EBL donned with optical transparency in the UV-VIS part of the spectrum, good mechanical properties and chemical stability.

The present study had, as a first step, the synthesis of appropriate CuO nanofillers. Subsequently, several CuO/PMMA polymer nanocomposite solutions were prepared and tested as positive tone EBL resists over silicon substrates. Contrast curve data and resolution analysis of the patterned CuO/PMMA films were used to evaluate the performance of the polymer nanocomposites as resist materials and the suitability of the suggested methodology as an alternative cost-efficient method for the production of CuO/PMMA nanocomposites. Within the framework of this study, four critical parameters were explored: the loading of the PMMA matrix with CuO nanofillers, the substrate coating conditions, the stability of the nanocomposite solutions over time and the role of the solvent comparing acetone to methyl ethyl ketone (MEK). The studies were complemented by Micro-X-ray fluorescence spectroscopy (μ-XRF) and a handheld XRF (hh-XRF), which were employed, to the best of our knowledge, for the first time to investigate the miscibility of the nanofillers within the polymer matrix and to assist in the selection of the optimum preparation conditions.

2. Materials and Methods

The study was conducted in two phases.

Phase 1 concentrated on synthesizing the appropriate CuO nanofillers and verifying the proof-of-concept of the suggested methodology for the production of CuO/PMMA polymer nanocomposites as EBL resists. A low-cost solution-based method was employed for the synthesis of the CuO nanofillers, because of its cost-efficiency and nanoparticle design versatility through simple key parameters, such as the temperature and the precursor concentration. After the synthesis of the appropriate nanofillers, three parameters were studied: the loading of the PMMA matrix with CuO nanofillers, the substrate coating conditions and the stability of the nanocomposite solutions over time.

Phase 2 focused on the role of the solvent; methyl ethyl ketone (MEK) was compared to acetone in terms of the suitability and performance of the CuO/PMMA nanocomposites as EBL resists.

2.1. CuO Nanofiller Synthesis

The CuO nanofillers were synthesized following a low-cost, wet chemical method, according to which copper acetate was hydrolyzed by sodium hydroxide (NaOH) in an aqueous solution followed by thermal decomposition. The chosen method was in essence a variation of the reduction of copper acetate with NaOH, as reported by Gupta et al. [34]. In brief, copper (II) acetate monohydrate (Sigma Aldrich/Merck KGaA, Darmstadt, Germany) was dissolved in DI water at room temperature to form a 65 mM solution. The solution was then placed on a hot plate and was heated under continuous magnetic stirring up to 80 °C. At that point, a 500 mM NaOH aqueous solution was added drop-wise in 2 mL doses until a copper acetate to NaOH molar ratio of 1:4 was reached. Upon addition of NaOH the translucent blue solution turned gradually opaque blue (Figure S1, Supplementary Materials, SM). The final solution was left under constant stirring at 80 °C for 2 h, during which a black sediment, characteristic of CuO synthesis, was formed and the solution turned transparent (Figure S1). The solution was left undisturbed to cool to room temperature overnight. Finally, the black precipitate was centrifuged (Kubota 2420, Kubota Corporation Tokyo, Japan) and washed with distilled water 3 times, and dried at 60 °C for 20 h and then at 90 °C for 24 h in an oven in presence of atmospheric air (Figure S1). The specific parameters of the synthesis (concentration, copper salt-to-NaOH molar ratio, temperature, duration of synthesis, etc.) were chosen after various combinations had been tested (Table S1). The final selection of the synthesis parameters was based on the requirement that they should lead to the formation of well-defined nanostructures of uniform average size (see Section 1 of SM for details, Figures S2 and S3 and Table S1). The nanopowder chosen for the production of the nanocomposites is shown in Figure 1. It consisted of almost spherical nanoparticles of pure CuO with an average diameter of ~10 nm.

Figure 1. SEM image of the CuO nanopowder selected to be used as the nanofiller for the PMMA composites. Magnification: ×150,000; Scale bar: 100 nm.

2.2. CuO Nanofiller Characterization

The CuO powders were structurally and morphologically characterized by Field-emission Scanning electron Microscopy (FE-SEM) with a JEOL JSM-7401f (Tokyo, Japan) and X-ray Diffraction (XRD) with q D500 SIEMENS Bragg-Brentano diffractometer, equipped with a pyrolytic graphite monochromator at a diffracted beam position and using Cu Kα radiation (CuKα1 Å: 1.54060, CuKa2 Å: 1.54439). The power conditions were set at 40 kV/35 mA, in addition to the aperture and the anti-scatter slit, which were set at 1°. The continuous step-scanning technique was used at steps of 0.03° with a measuring time of 2 s/step and the recorded 2θ range was from 2.0° to 100.0°. The XRD results are summarized in Figure S2.

2.3. PMMA Preparation

5% w/w PMMA in propylene glycol monomethyl ether acetate (PGMEA) solutions were prepared for Phase 1 and 6% w/w in PGMEA were prepared for Phase 2 using PMMA with a molecular weight MW = 996 k from Sigma Aldrich. Dissolution of PMMA was aided by the use of a magnetic stirrer in conjunction with low thermal plate heating (<70 °C) for 72 h.

2.4. PMMA/CuO Polymer Nanocomposite Solutions

In order to form the polymer nanocomposites, the following method was used in both phases: CuO nanopowder was added to a pre-calculated volume of acetone and the solution was vigorously stirred on a magnetic stirrer for 30 min at 40 °C in an effort to break up as many agglomerates as possible. A specified amount of acetone was added into the prepared PMMA/PGMEA solution prior to the addition of the nanofillers and the new solution was stirred at 40 °C for 30 min. Subsequently, the two mixtures were combined, so that after the addition of the acetone-CuO solution, a 4% w/v PMMA solution was obtained, while the loading of CuO nanofillers was 1% w/v, 2% w/v or 3% w/v. The final CuO/PMMA solution was stirred for another 30 min at 40 °C to improve homogenization (Figure S5a). Additional CuO/PMMA solutions were prepared, in which a small amount of deflocculant was added (Darvan C, Vanderbilt Minerals, LLC, Norwalk, CT, USA) to study its effect on the stability and homogenization of the polymer nanocomposite solutions. Darvan C is an ammonium salt of poly (methacrylic acid) with an average molecular weight of 10,000–16,000 g/mol and is commercially available as an aqueous solution with an active content 25%. Finally, a 4% w/v PMMA-acetone solution without any nanofillers was prepared to be used as a reference, hereafter referred to as REF. In Phase 2, the same preparation procedure was followed with the concentration of the nanofillers kept fixed at 1% w/v. Two different solvents were studied, acetone and methyl ethyl ketone (MEK), so as to evaluate the role of the solvent in the nanocomposite preparation. Three different solutions were prepared for each solvent; one containing no nanofillers used as a reference, one with only CuO nanofillers and one with CuO nanofillers and a small amount of deflocculant (Darvan C).

2.5. Electron Beam Lithography

The CuO/PMMA nanocomposites were tested as positive tone EBL resists according to the following procedure. All CuO/PMMA solutions were spin-coated onto 2.5 × 2.5 cm^2 Si pieces obtained after dicing 3" Si wafers. The silicon substrates were thoroughly cleaned prior to the deposition with organic solvents and a piranha solution. Three different rotation speeds were tested during Phase 1, namely 1000 rpm, 3000 rpm and 4000 rpm (30 s) in order to select the most appropriate spin-coating conditions. Three different nanofiller loadings were tested, namely 1%, 2% and 3% w/v. Additionally, the addition of deflocculant was examined. All the samples underwent a post apply bake (PAB) at 180 °C for 1 min. For Phase 1, the coated samples were named after the CuO/PMMA solution used as follows: "X%CuO-Yk", where X was the concentration of CuO (1%, 2% or 3% w/v) and Y was the spin coating speed (1k, 3k or 4k corresponding to 1000 rpm, 3000 rpm and 4000 rpm, respectively)."REF-Yk" corresponds to the reference samples prepared by the reference 4% w/v PMMA-acetone solution without any nanofillers spin-coated at Yk rpm. In the case of the deflocculant addition, the sample name contains the ending "-DF". All samples that were studied in Phase 1 are listed in Table 1. The reported thicknesses of the films were determined by stylus profilometry (Ambios XP2, Ambios Technology, Inc, Milpitas, CA, USA) after development.

Table 1. Phase 1 samples coated with CuO/PMMA polymer nanocomposite solutions. REF-Yk correspond to films prepared by the 4% w/v PMMA-acetone solution without any nanofillers spin-coated at Y krpm. X%CuO-Yk correspond to films prepared containing X% w/v CuO nanofillers spin-coated at Y krpm. DF denotes the addition of deflocculant.

Sample Name	CuO Concentration (w/v)	Deflocculant	Spin Coating Speed (rpm)	Thickness (nm)
REF-1k	0%	NO	1000	573
REF-3k	0%	NO	3000	340
REF-4k	0%	NO	4000	272
1%CuO-1k	1%	NO	1000	601
1%CuO-3k	1%	NO	3000	336
1%CuO-4k	1%	NO	4000	293
2%CuO-1k	2%	NO	1000	570
3%CuO-1k	3%	NO	1000	N/A
1%CuO-1k-DF	1%	YES	1000	616

During Phase 2, the spin coating speed was fixed to 1000 rpm and the CuO loading to 1% w/v, since the main goal was to examine the effect of the solvent. Two series of samples were prepared, one for acetone and one for MEK. Each series contained one reference film produced with the "bare" PMMA solution, one film created with the 1% CuO/PMMA solution, one with the 1% CuO/PMMA solution containing the deflocculant and one with the 1% CuO/PMMA solution without deflocculant being filtered during the drop-casting using PTFE filters with 0.2 μm pores (Machery-Nagel GmbH & Co, Dueren, Germany). The thickness of the films was determined both via ellipsometry (M2000-F, J.A. Woollam Co., Lincoln, NE, USA) after PAB and via stylus profilometry after development.

For Phase 2, the samples were named as follows "X-REF", "X-CuO", "X-DF" and "X-FIL", where X denotes the solvent used (X: ACE for acetone; MEK for MEK), "REF" corresponds to the reference solutions without nanofiller, "CuO" denotes that only nanofillers were added, "DF" denotes the addition of deflocculant and "FIL" denotes the filtering procedure during drop-casting. The samples of Phase 2 are summarized in Table 2.

Table 2. Phase 2 samples coated with CuO/PMMA polymer nanocomposite solutions. Prefix ACE corresponds to solutions prepared with acetone as the solvent, while prefix MEK corresponds to solutions prepared with MEK as the solvent. DF denotes the addition of deflocculant and FIL denotes that the solution was filtered.

Sample Name	CuO Loading	Solvent	Deflocculant	Filtering
ACE-REF	0%	Acetone	NO	NO
ACE-CuO	1%	Acetone	NO	NO
ACE-DF	1%	Acetone	YES	NO
ACE-FIL	1%	Acetone	NO	YES
MEK-REF	0%	MEK	NO	NO
MEK-CuO	1%	MEK	NO	NO
MEK-DF	1%	MEK	YES	NO
MEK-FIL	1%	MEK	NO	YES

For all samples in both phases, contrast curve patterning was conducted using a Raith EBPG5000+ e-beam writer (Raith GmbH, Dortmund, Germany) operating at 100 keV, in order to compare and characterize resist formulations via their response to exposure dose. Microscale structures (200 µm-wide squares, Figure S4a) intended for contrast curve data acquisition were exposed without the proximity effect correction at a detailed set of exposure doses (50–645 µC/cm^2), with a 15 µC/cm^2 step. Resolution patterns were exposed at a range of base doses (330–700 µC/cm^2), below and above the observed dose to clear, with a 40 µC/cm^2 dose step. 200 µm-long rectangular ribbons of variable width (300 nm, 500 nm, 1 µm, 5 µm, 10 µm, and 20 µm) were defined both by direct exposure (grooves) and by exposure of their periphery (protruding ridges), in order to probe the ability to design and transfer patterns onto the nanocomposite/resist films under different exposure conditions destined for different applications and architectures (see the schematic representation Figure S4b,c). Patterns for resolution studies were designed on KLayout and lithographic data preparation, including proximity effect correction, was performed using Beamer from GenISys. EBL was conducted using a 30 nA e-beam current and beam shot pitch was set to 25 nm. The development duration was set to 60 s. A 7:3 isopropanol/DI water co-developer solution was used for the development of samples followed by isopropanol rinse and N$_2$ blow.

Contrast curve data (remaining film thickness in the exposed area) were acquired via stylus profilometry. Contrast curves measure the resist formulations' sigmoidal response to exposure dose, while contrast (γ) is a dimensionless parameter that measures the films' characteristic ability to conform to dose variations, under particular processing conditions. Typically, γ is extracted from the linear portion of the curve close to zero thickness, however, in our analysis, calculation of γ values is based on best curve fit, using the Ziger-Mack methodology [35]. The resist formulations' performance, in terms of resolution, was assessed via optical microscopy and qualitative inspection of e-beam defined resolution patterns. Structural response to exposure dose bears information on the limitations and capabilities of resist variations and is discussed in the Results section.

2.6. Micro X-ray Fluorescence (µ-XRF) and Handheld-XRF Characterization (hh-XRF)

The CuO/PMMA-coated Si substrates were also characterized via µ-XRF in order to determine whether the CuO nanofillers were homogeneously dispersed within the PMMA matrix or whether they only form agglomerates, as seen through optical and electron microscopy. Such an approach, to the best of our knowledge, has not been attempted before.

The µ-XRF spectrometer probe used in this work consists of a micro focus Rh-anode tube, a polycapillary X-ray lens as a focusing optical element (IfG-Institute for Scientific

Instruments GmbH, Berlin, Germany), with a focal distance equal to 21.2 mm, and a nominal gain factor that varies between 3625–4900–1200 for energies within the intervals of 3–5, 10–15 and 25–30 keV, respectively. The X-ray detection channel consists of an electro-thermally cooled 10 mm^2 silicon drift detector (X-Flash,1000 B) with full width at half maximum at 5.89 keV equal to 146 eV at 10 kcps coupled with a digital signal processor. Three different stepping motors, coupled with the spectrometer head, allows for its three-dimensional movement, facilitating the elemental mapping studies. Finally, a color charge-coupled device camera (x13), a dimmable white light-emitting diode for sample illumination and a laser spot assist in the documentation and sample alignment. The spectrometer spatial resolution (FWHM) for the excitation of the Cu-Kα line was measured to be ~80 µm. For Cu quantification purposes, the µ-XRF spectrometer was calibrated by means of a multi-elemental, nm-scaled sample manufactured by AXO-Dresden, GmbH. The stratified sample was composed of ~10 nm individual layer thicknesses of Cr, Al, Ni, Cu and Ti elements deposited on a few micrometers polymer film. This reference sample belongs to the same batch of similar reference materials developed for synchrotron radiation experiments [36]. The X-ray tube measurement conditions were set at 50 kV, 600 µA using an unfiltered exciting beam, whereas the µ-XRF scanning parameters were set as follows: step size 0.1 mm, 20–25 s measurement time per step with a typical investigated sample area of about 1.5 × 1.5 mm^2. For certain samples, a larger area was scanned. The spectrum deconvolution and quantification were carried out using the PyMca analysis software [37]. Si substrates coated with the 4% w/v PMMA solutions at all 3 rotation speeds were used as references.

In addition to the scanning µ-XRF measurements, a hh-XRF analyzer with Rh anode transmission X-ray tube (Tracer 5i, Bruker) was used to perform selected screening measurements of the average Cu deposited areal density. The hh-XRF measurements were performed at 30 kV/110 µA high voltage/current operating conditions using the combined Ti/Al filter for the exciting X-ray beam provided by the manufacturer.

3. Results and Discussion

3.1. Phase 1

As a first step, the EBL patterned samples were examined under an optical microscope (Figure 2). It was readily observed that increasing the concentration of CuO to 2% or 3% w/v resulted in the formation of large agglomerates that reached up to 100 µm (Figure 2c,d) in size, rendering these concentrations unsuitable for the production of CuO/PMMA nanocomposites. The agglomerates were so large that they prevented obtaining accurate and reliable stylus profilometry measurements necessary for the determination of resist thickness. In addition, it was also observed that the agglomerates were not successfully removed from the exposed areas after the EBL development step. For those reasons, samples 2%CuO-1k and 3%CuO-1k were not included in the subsequent analysis of the films with respect to their performance as resists and were deemed unsuitable for any practical application of the nanocomposites. However, it is worth noting that it was still possible to pattern the nanocomposite films with e-beam lithography regardless of the agglomerates suggesting that it might still be possible to use solutions of higher nanofiller loadings after filtration. This aspect has not been addressed in this work and is still under investigation.

In contrast, when the concentration of the nanofillers was kept to 1% of the agglomerates were significantly smaller in size (Figure 2b,c,f,g), but still their size, as shown after SEM inspection, could reach up to 1–3 µm (Figure 3a,b). The scattered agglomerates would protrude through the film (Figure 3b), demonstrating the need to find a way to hinder their formation. The rest of the film exhibited great homogeneity (Figure 3c), but it was not possible to determine, through SEM, whether individual nanofillers were dispersed within the film.

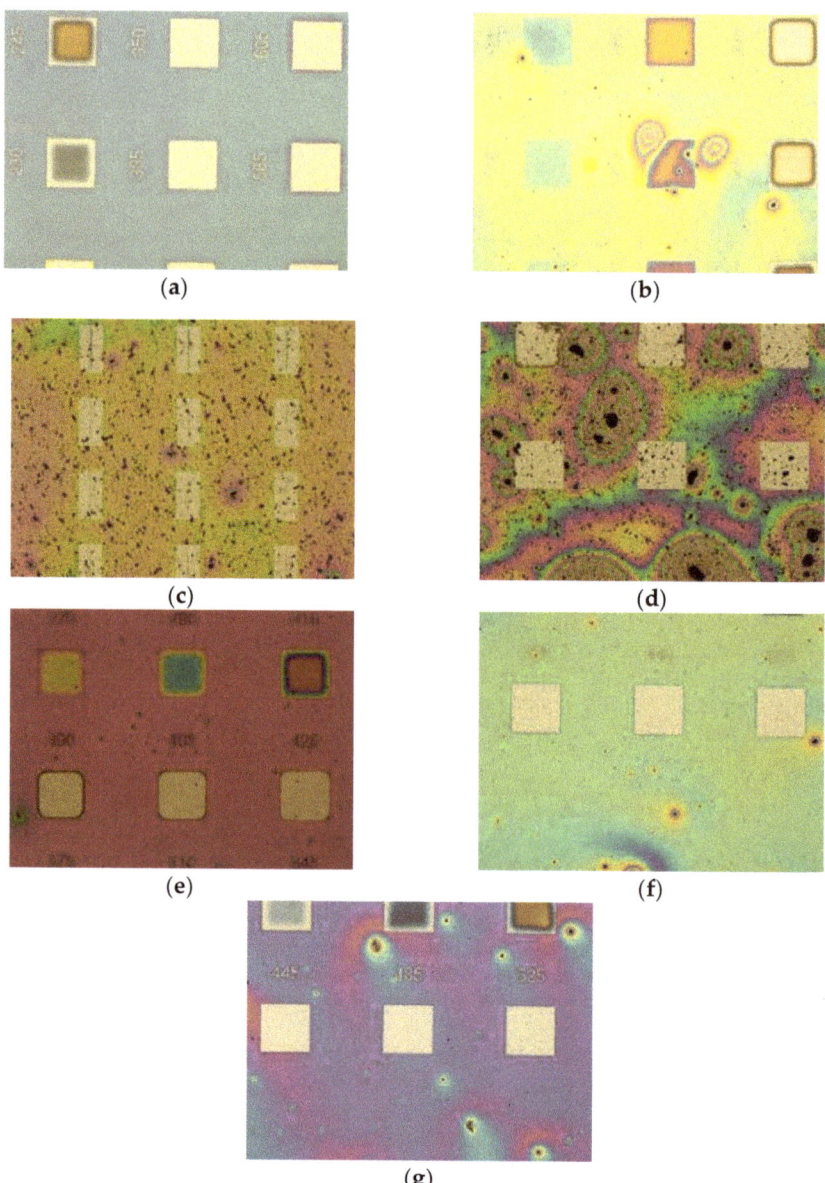

Figure 2. Optical microscope images (magnification: ×10) from the contrast curve patterns used to determine the suitability of the CuO/PMMA nanocomposites as resist materials. The square patterns have a size of 200 μm × 200 μm (detailed description can be found in Section 2 of SM). Analytically, (**a**) REF-1k, (**b**) 1%CuO-1k, (**c**) 2%CuO-1k (picture of resolution structures to demonstrate their characteristic pattern), (**d**) 3%CuO-1k, (**e**) 1%CuO-1k-DF, (**f**) 1%CuO-3k, and (**g**) 1%CuO-4k.

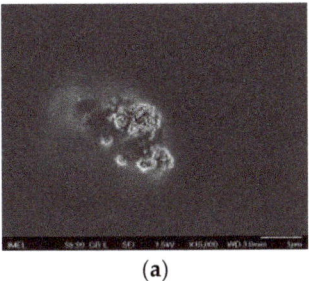 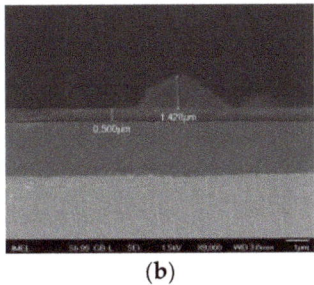 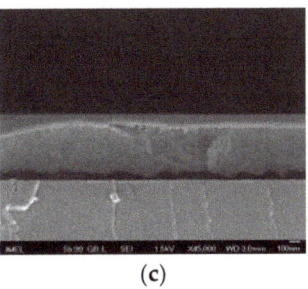

(a) (b) (c)

Figure 3. SEM images of sample 1%CuO-1k (**a**) top-down view showing a typical agglomeration of nanofillers found in the film, (**b**) cross-section of one of the nanofiller agglomerates demonstrating their relative size compared to the nanocomposite film thickness and (**c**) cross-section and magnification of the nanocomposite film at a section devoid of agglomerates. Scale bars: 1 μm in (**a**,**b**); 100 nm in (**c**).

During the experiments, it was observed that, after approximately 24 h, the nanofillers would sediment in all CuO/PMMA solutions (Figure S5b). The inability to form stable inorganic/polymer solutions is one of the main challenges of polymer nanocomposites. Therefore, in order to address this issue and to find a solution, which would still keep the suggested methodology as simple and as cost-efficiently as possible, a deflocculant was used to examine whether it would extend to the stability of the solutions. The chosen deflocculant was Darvan C, a substance commonly employed in ceramic dispersions giving low viscosity slip and low foam production [38–40]. Upon the addition of Darvan C, the solution turned into a stable emulsion and no signs of sedimentation were observed for several days. Over the course of a week, light blue sediments appeared on the vial walls, an indication that copper (II) hydroxide salts were formed due to the presence of the amine groups of Darvan C (Figure S5c). After a month, most of the copper oxide had turned into copper (II) hydroxide dehydrate, as attested by the light-blue co-aggulated sediment and the clear color of the solution, similar to that of pure PMMA (Figure S5d). This suggests that, even though immediate sedimentation was prevented, the CuO/PMMA/DF nanocomposites have a shelf-life of approximately 1 week, which is still an improvement to the limited shelf-life of 24 h in the absence of deflocculant. As far as the film is concerned, it still contained agglomerates that were visible through the optical microscope (Figure 2e), which, however, were more uniform in size and more homogeneously dispersed compared to all the other films.

Resist variations were evaluated in terms of their bulk lithographic properties (thickness, sensitivity, and resolution) via contrast curve data analysis and optical inspection. Spin curves (resist thickness after development versus spin speed) were constructed from profilometric measurements of the contrast curve patterns.

As a first observation, it was seen that the films produced, both the reference and nanocomposite ones, were considerably thicker compared to films produced under the same conditions by 4% w/w PMMA in PGMEA, as described in Section 2.3 (even though the authors acknowledge that 4% w/w PMMA in PGMEA is not exactly the same as a 4% w/v PMMA/PGMEA/Acetone solution, it is the closest in terms of a reference resist). 4% w/w PMMA in PGMEA solutions typically provide films with a thickness of 260 nm, 150 nm and 130 nm after PAB, when resist-spinning is performed at 1k, 3k and 4k rpm, respectively (Figure 4). It is postulated that acetone, as a more volatile solvent with respect to PGMEA, evaporates much faster during the spin-coating process, resulting in thicker films. Moreover, the addition of a different solvent has a direct effect on the viscosity, which, in this case, seems to be increased. Turning the focus on the nanocomposite films, the addition of nanofillers with 1% w/v loading slightly increased the thickness of produced films (Figure 4). The presence of deflocculant even further increased the thickness of the nanocomposite film suggesting, in a very indirect way, that its presence might have assisted in the nanofiller distribution within the polymer matrix.

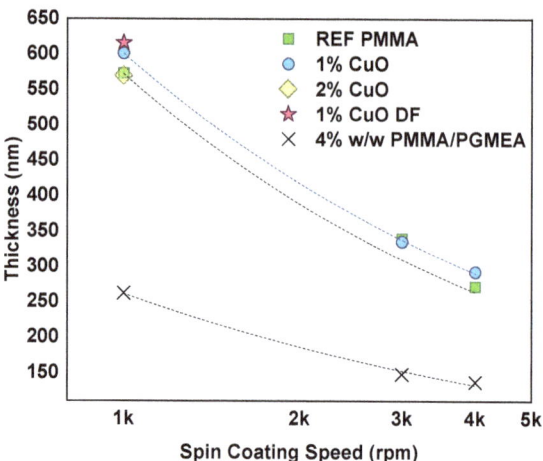

Figure 4. Spin curves for the "bare" PMMA reference samples (green squares), the nanocomposites containing 1% w/v CuO (blue circles), sample 2%CuO-1k (yellow diamond) and 1%CuO-1k-DF (magenta star). Crosses correspond to the 4% w/w PMMA/PGMEA.

In an effort to assess whether the films contained dispersed nanofillers and not only agglomerates, and to substantiate the validity of the assumptions above, the samples were investigated by scanning μ-XRF measurements. The pixel Cu-Kα μ-XRF data are summarized in Figure 5 and have all been deduced by means of the PyMca software accounting for rather negligible blank contributions. As readily seen, the films with 2% and 3% w/v CuO loadings yielded signals that were two orders of magnitude larger than all samples of 1% w/v loading. The signals also varied by two orders of magnitude from point to point across the samples, a fact that it is in accordance with the presence of the large agglomerates, as well as the smaller agglomerates that spot the surface observed through optical microscopy (Figure 2c,d). All the samples with 1% w/v nanofiller loading exhibited more or less the same behavior. A sizeable percentage (10−25%) of the signals were below the limit of detection (LOD), the majority of the signals (85−70%, respectively) were between the LOD and the limit of quantification (LOQ) was set as 3*LOD, and only 5% were above LOQ, most likely related to the presence of large agglomerates. However, the signals also spanned across two orders of magnitude, suggesting that individual nanofillers were most likely dispersed within the volume of the polymer matrix, but their small size and spatial distribution results into low signals. It should be noted that sample 1%CuO-3k yielded results below or very close to the LOD and was not further assessed with this particular method.

Despite the fact that most of the signals were below the LOQ, an attempt was made to calculate the areal density (Figure 5b; Table 3). It should be clarified that the Cu areal densities are reported for the whole scanned area by summing all the individual pixel (0.1 mm) size measurements. The LoD for Cu areal density, as obtained from the sum spectrum, is in the order of 0.02 μg/cm^2 corresponding to a measurement time of about 5000 s. The theoretical calculation for the expected Cu areal density of the films taking into account their thickness and nanofiller loadings would be in the order of 0.2–0.9 μg/cm^2 (values given in parentheses in Table 3). For the 1% w/v nanofiller loadings, the experimental values were within the same order of magnitude with the theoretical expectations, but for the 2% w/v loading, the measured areal density was increased by two orders of magnitude, corroborating with the scenario of aggregate formation and their inhomogeneous dispersion within the polymer.

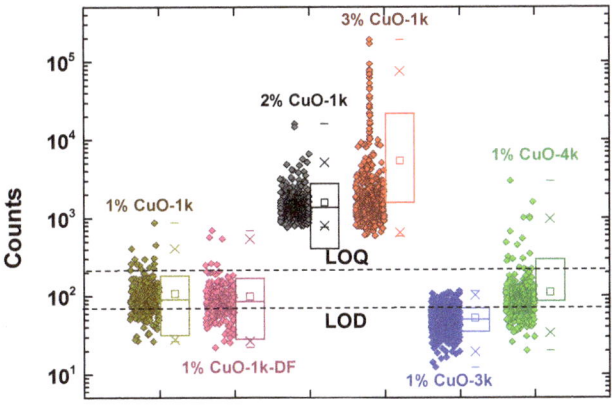

(a)

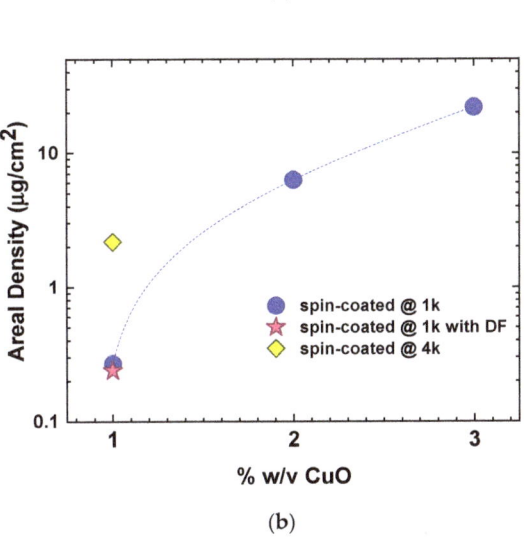

(b)

Figure 5. Synopsis of μXRF measurements for all CuO/PMMA samples (Phase 1). (a) Scattered solid diamonds depict the intensity signal from each point (pixel), while open squares show the average signal value. The large rectangles show the range between 25% and 75% of the maximum obtained signal, the crosses mark the range between 1% and 99% of the maximum obtained signal and the horizontal dashes show the minimum and maximum obtained values. Dashed lines show the limit of detection (LOD) and the limit of quantification (LOQ) set as 3*LOD. (b) Cu areal density as a function of the nanofiller loading and the preparation parameters. Blue circles: films of various loadings spin-coated at 1 krpm; Magenta star: sample 1%CuO-1k-DF with 1% CuO nanofillers and deflocculant spin-coated at 1 krpm; Yellow diamond: sample 1%CuO-4k with 1% CuO nanofillers and spin-coated at 4 krpm.

Table 3. Summary of Phase 1 Results.

Sample Name	Film Thickness (nm)	Dose-to-Clear (μC/cm^2)	γ	XRF Areal Density (μgr/cm^2)
REF-1k	575	420	2.89	N/A [a]
REF-3k	340	350	2.48	N/A
REF-4k	270	345	2.53	N/A
1%CuO-1k	600	485	2.99	0.27 (0.48) [b]
1%CuO-3k	335	410	2.85	−(0.26)
1%CuO-4k	290	425	2.95	0.30 (0.23)
1%CuO-1k-DF	615	420	2.72	0.24 (0.49)
2%CuO-1k	570	N/A	N/A	6.3 (0.91)
3%CuO-4k	N/A	N/A	N/A	22.1 (-)

[a] N/A: non-applicable. [b] Values in parentheses are the theoretical calculations of the Cu areal density.

As a final step of this phase, the contrast patterns of the films were evaluated. The behavior of the nanocomposite films is illustrated in Figure 6a,b. Sensitivity (expressed as a dose to fully clear the resist film, including remains at the corners of the exposed square) and contrast γ are demonstrated in Figure 6c,d, as a function of the films' thickness and preparation conditions. The results are summarized as dose-to-clear (DTC) versus contrast γ in Figure 6e, since the reference and their respective nanocomposite films had comparable thicknesses (Figure 4). For the reference samples, DTC was between 345 μC/cm^2 and 420 μC/cm^2, while upon addition of the nanofillers, DTC appreciably increased, ranging from 425 μC/cm^2 to 485 μC/cm^2, indicating that the presence of the CuO nanofillers affect the sensitivity of the resist matrix. Contrast γ values are within the range of 2.5–2.9 for all the samples. For the "standard" 4% w/w PMMA/PGMEA, the associated DTC for resist films of ~260 nm (spin-coated at 1k rpm) is in an average 360 μC/cm^2 and contrast γ is ~2.9 (Figure 6e), when EBL is performed at 100 kV and the development duration is set to 60 s.

It appears that the addition of acetone to PMMA/PGMEA (REF samples) slightly decreases contrast γ without affecting DTC. According to Gaikwad et al. [41], acetone is a relatively good solvent for PMMA, and in this work, it did not drastically affect the properties of PMMA as a positive tone resist in terms of contrast γ and DTC. Upon addition of nanofillers, DTC increased significantly, while contrast γ returned to the value of 2.9–3.0. The increase of DTC is attributed to scattering of the beam electrons by the nanofillers and serves as an indirect verification that the metal oxide nanoparticles, even though not directly observable by means of microscopy, are indeed dispersed within the polymer matrix. In contrast to the results in References [16,17], according to which the presence of 1 nm silica nanoparticles in ZEP520® increased the resolution of the resist without sacrificing the sensitivity and contrast, the presence of the CuO nanofillers in PMMA in this work decreases the sensitivity, but slightly increased the contrast with respect to the reference samples (Figure 6e). This can be attributed to the larger size of the CuO nanofillers (~10 nm), which can scatter the electrons more efficiently than the 1 nm-wide silica nanoparticles. It is also suggested that the presence of agglomerates might also play an additional role in electron scattering. The slight increase of γ with respect to the reference samples might also be due to the role of the nanofillers as scattering centers, which do not just reduce the penetration depth of the incident electrons (increasing thus DTC), but also their lateral range and the spreading of secondary electrons [16].

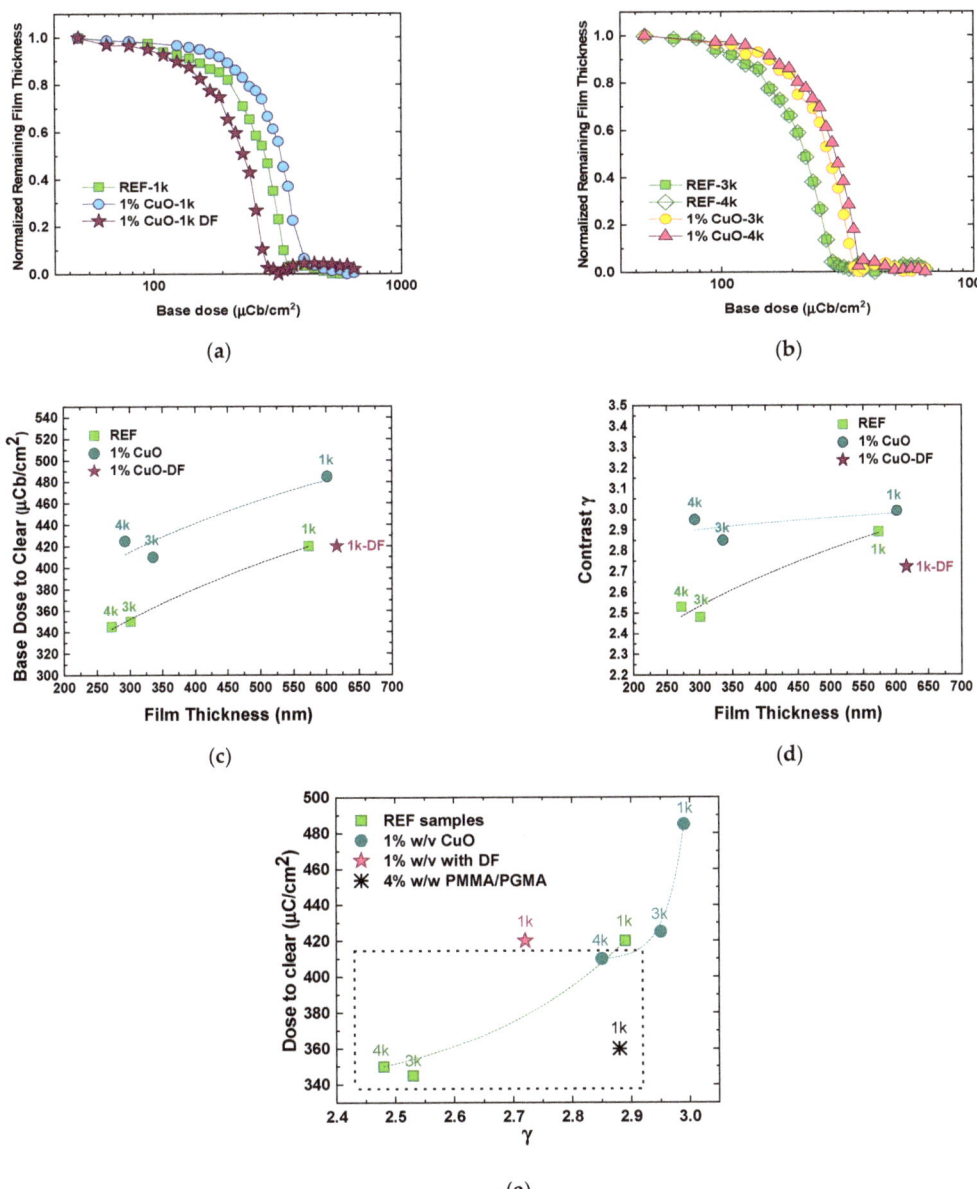

Figure 6. Contrast curves (Normalized Remaining Thickness as a function of the base dose) for patterned samples with films spin-coated at (**a**) 1000 rpm, (**b**) 3000 and 4000 rpm; (**c**) sensitivity (expressed as the minimum DTC) versus film thickness, (**d**) contrast γ versus film thickness, and (**e**) dose-to-clear versus contrast γ for all nanocomposite films of Phase 1 that were evaluated as EBL resists. The dotted rectangle includes resist films of similar thicknesses of ~260–300 nm. Dashed lines in graphs of (**c**–**e**) serve only as a guide to the eye, while labels indicate the spin-coating rotation speed in rpm.

Notably, the addition of the deflocculant "returned" the behavior of the nanocomposite closer to that of bare PMMA in terms of DTC. This behavior is suggestive that the deflocculant, apart from the prolonged shelf-life of the CuO/PMMA solutions, may offer a better homogenization of the nanocomposite and a more uniform dispersion of the nanofillers within the PMMA matrix. If the nanofillers are more uniformly dispersed (as already suggested by optical microscopy) scattering of the electrons may not be as profound as in the case of sample 1%CuO-1k, which is "dotted" with larger agglomerates and the film as a resist requires approximately the same base dose to be fully cleared as the bare PMMA.

The findings of Phase 1 are summarized in Table 3.

Phase 1 proved that the present methodology, despite its simplicity, can be used to produce CuO/PMMA nanocomposites as positive tone EBL resist materials. The results suggested that nanofiller loadings should not exceed 1% w/v, while it is important to more efficiently control the formation of metal oxide aggregates. The use of the deflocculant significantly extended the shelf-life of the CuO/PMMA and offered an improved homogenization and nanofiller dispersion within the polymer matrix. These results led to Phase 2, which concentrated on examining four parameters:

(1) In order to improve the nanofiller dispersion, the volume of the additional solvent was increased. For that reason, the initial PMMA/PGMEA concentration was slightly increased, from 5% w/w to 6% w/w, to allow for a larger amount of solvent to be added to reach the final 4% w/v PMMA concentration in the CuO/PMMA solution.
(2) One more solvent, MEK, was tested in conjunction to acetone to test the role of the solvent
(3) The effect of the deflocculant was re-examined with the new conditions of increased solvent volume and the new solvent
(4) The effect of filtration prior to spin-coating deposition of the resist films in the absence of DF was tested.

3.2. Phase 2

As already described in Section 2 and summarized in Table 2, eight resist films were prepared and patterned by EBL, four for each one of the two solvents. The CuO loading was set to 1% w/v and the PMMA concentration to 4% w/v, both with respect to the final solution volume. The spin-coating rotation speed was fixed to 1000 rpm. XRF measurements, EBL patterning and subsequent contrast curve and resolution analysis was conducted in exactly the same manner as in Phase 1. The contrast patterns of the samples are shown in Figures S6 and S7, while the resolution patterns are shown in Figures S8 and S9. All data compiled from Phase 2 are summarized in Table 4.

Table 4. Summary of Phase 2 results.

Sample Name	Film Thickness [a] (nm)	Dose-to-Clear ($\mu C/cm^2$)	γ	Resolution [b] (Groove/Ridge)	XRF Areal Density ($\mu gr/cm^2$) [e]
ACE-REF	530/485	460	3.4	300 nm/300 nm	N/A
ACE-CuO	510/460	405	N/A	300 nm/10 μm	2.2/8.7 (0.37)
ACE-DF	-/465	405	3.1	300 nm/5 μm [c]	2.7/2.0 (0.37)
ACE-FIL	575/503	435	3.8	300 nm/10 μm [c]	-/0.02 (0.40)
MEK-REF	495/450	445	3.6	300 nm/300 nm	N/A
MEK-CuO	420/420	390	N/A	300 nm/-	25.6/19.3 (0.34)
MEK-DF	395/395	390	3.3	300 nm/5 μm [d]	0.46/2.4 (0.32)
MEK-FIL	455/420	410	3.6	300 nm/5 μm [d]	-/0.04 (0.34)

[a] First value corresponds to ellipsometry result after PAB; second value to stylus profilometry of contrast patterns after development. [b] The first number indicates the minimum feature size achieved for the formation of grooves, while the second for protruding ridges. [c] Recorded resolution achievable only for moderate doses close to the minimum dose-to-clear–High sensitivity to overdose. [d] Recorded resolution achievable only for dose to clear—higher doses result in no ridges. [e] First value obtained by μ-XRF/ Second value obtained by the hh-XRF analyzer; Values in parentheses are the theoretical calculations of the Cu areal density.

Contrary to what was expected, the increased amount of solvent did not ameliorate the nanofiller dispersion, but instead resulted in the formation of more and larger aggregates. Both samples, ACE-CuO and MEK-CuO, could not be fully assessed in terms of their sensitivity due to the large number of aggregates that did not allow us to obtain reliable stylus profilometry measurements for all base doses and construct a contrast curve that could be analyzed. The data collected have only been used to calculate the film thickness and standard deviation after the EBL, but not to calculate the sensitivity of the film. Nonetheless, the resolution patterns were inspected to form a more complete picture of their behavior. The reason behind the formation of numerous aggregates with the increase of the solvent volume is still under investigation, but an initial assumption is that the lower viscosity does not enhance the miscibility, but instead promotes phase separation of the constituent materials.

The film thicknesses, as measured by ellipsometry, with respect to the preparation conditions (before EBL exposure), are presented in Figure 7 alongside the average film thickness and its standard deviation, as measured from the stylus profilometry after development of the film. All films were thinner with respect to the Phase 1 samples, which means that the viscosity was reduced. The acetone-based films were, in all cases, slightly thicker than the MEK-based ones. When MEK was employed, the addition of the deflocculant resulted in the thinnest films. The filtration resulted in films comparable in thickness to the reference ones, regardless of the solvent.

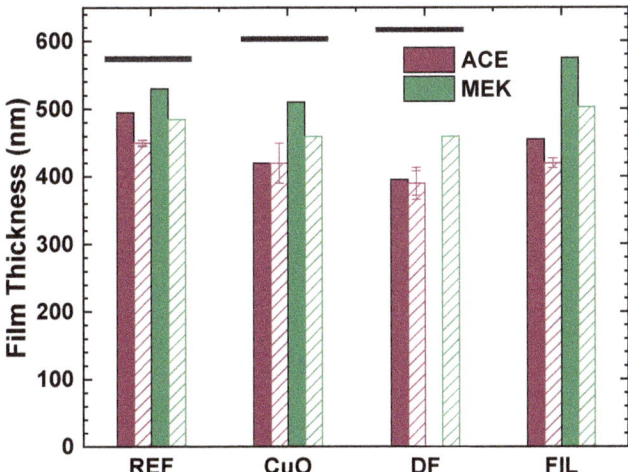

Figure 7. Film thickness of Phase 2 CuO/PMMA nanocomposite films, as measured by ellipsometry prior to EBL (solid bars) and as calculated by stylus profilometry of the contrast patterns after development (striped bars). Magenta bars correspond to acetone-based nanocomposites; green bars correspond to MEK-based nanocomposites. Horizontal black lines denote the film thickness of the respective samples from Phase 1.

The μ-XRF measurements corroborated the fact that increasing the amount of solvent resulted in an increased number of aggregates, as well as aggregates of larger size (Figure 8a). When the results of ACE-CuO and MEK-CuO are compared to the corresponding results of 1%CuO-1k, it is readily seen that there was a significant increase in the signal intensity dispersion over three orders of magnitude with more than 50% of the signals being over the LOQ affirming the formation of larger aggregates. When a defloculant was added the signal dispersion was decreased and most of the signals were in the same order of magnitude, again corroborating the optical microscopy and profilometry observation—as well as the results of Phase 1—that the presence of the deflocculant limits the formation of aggregates and improves the miscibility of the nanofillers without completely eliminating the aggregate presence. When comparing ACE-DF and MEK-DF to 1%CuO-1k-DF, it was again observed that the increase in the solvent content resulted in an increase in aggregate number and size. When filtration was applied, the μ-XRF signals dropped close to or even below the LOD of the method and it was not possible to affirm the presence of nanofillers within the resist film or to safely deduce any conclusion. For that reason, additional measurements were performed with the hh-XRF analyzer, offering the possibility to deduce average Cu areal densities over a large irradiated area (approximately described as a circle with 8 mm in diameter, Figure 8b). In fact, the hh-XRF analysis of the ACE-FIL sample succeeded to detect a minimum, but not quantifiable amount of Cu above the respective LOD. However, the non-consistent Cu areal densities deduced by the μ-XRF and hh-XRF spectrometers (ACE-CuO and MEK-DF in Figure 8b) might be due to the inhomogeneity of the deposited area, as the μ-XRF analyzed ~2.25 mm^2 versus the ~47.1 mm^2 probed by the hh-XRF analyzer. Comparing the theoretical values to the experimental values, one readily observes that the Cu areal density exceeds by one or two orders of magnitude the anticipated areal density that would correspond to the nominal nanofiller loading of 1% w/v (~0.3–0.4 μgr/cm^2), implying that the nanofillers have aggregated into larger structures non-homogeneously dispersed within the solution. Applying filtration resulted in areal densities very close to the LOD and one order of magnitude lower than the one corresponding to 1% w/v, suggesting that the aggregates were retained in the filter and only a smaller quantity than 1% w/v of nanofillers were present in the resist film.

The contrast curves and subsequent sensitivity and contrast γ analysis was performed based on the thickness measured after development. The analysis (Figure 9) demonstrated that, when the CuO nanofillers are added to the PMMA, the DTC drops with respect to the bare PMMA, as was also observed in Phase 1. This further adds to the scenario that the electrons scatter on the nanofillers. This is further substantiated by the fact that the lowest DTCs are observed for the nanocomposites containing the deflocculant; despite the improved homogenization and stability, the deflocculant cannot prevent the formation of agglomerates, which act as large scattering centers. On the contrary, when the CuO/PMMA solution is filtered most of the agglomerates do not end up in the film, and the dose-to-clear increases with respect to the nanocomposites with deflocculant. Still, the presence of nanofillers and the related electron scattering maintain DTC levels to lower values with respect to the reference films (Figure 6c–e). When comparing the two solvents, the dose-to-clear is in general lower for the case of MEK, a fact that is expected, given that the resulting resist films are thinner to begin with.

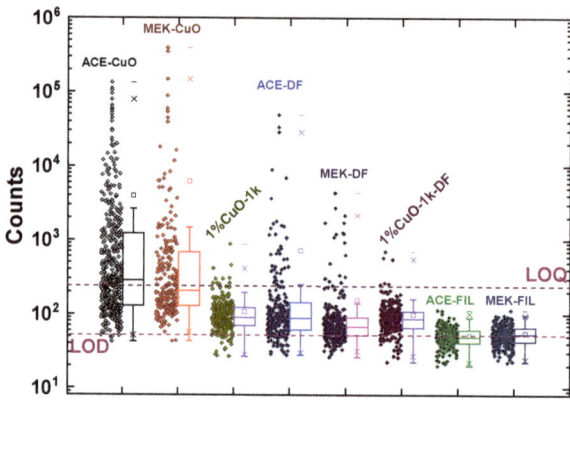

(a)

Figure 8. *Cont.*

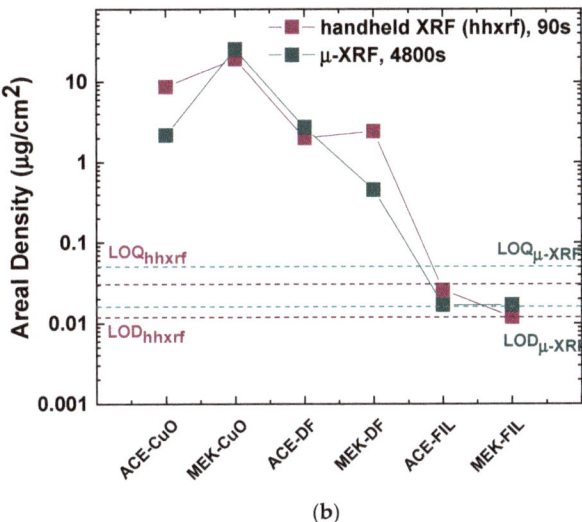

(b)

Figure 8. (a) Synopsis of μ-XRF measurements for all Phase 2 CuO/PMMA samples. The corresponding measurements of 1%CuO-1k and 1%CuO-1k-DF from Phase 1 have been included for comparison. Scattered solid diamonds depict the intensity signal from each point, while open squares show the average signal value. The large rectangles show the range between 25% and 75% of the maximum obtained signal, the crosses mark the range between 1% and 99% of the maximum obtained signal and the horizontal dashes the minimum and maximum obtained values. Dashed lines show the limit of detection (LOD) and the limit of quantification (LOQ) set as 3*LOD. (b) Comparison of the calculated Cu areal density by the hh-XRF analyzer (magenta bars) and the μ-XRF spectrometer system (green bars). Dashed lines denote the LOD and LOQ determined for measurement times of 4800 s and 90 s for the μ-XRF and hh-XRF spectrometers, respectively.

As far as contrast γ is concerned, in accordance with the results of Phase 1, the addition of the deflocculant decreases its value compared to all other samples, irrespective of the solvent used. Nonetheless, for the case of MEK, all resists had γ values that did not differ considerably among them (Figure 6d,e). In contrast, the γ values for the acetone-based films had considerable variations among them. Additionally, γ for the acetone-based resists was increased compared to Phase 1 samples. Notably, γ between ACE-DF and ACE-FIL differed by 0.7 (an increase of almost 25%).

The resolution patterns of acetone and MEK based CuO/PMMA nanocomposite films (Figures S8 and S9) show that the addition of the nanoparticles have an observable influence on the resolution of the protruding ridges, while the resolution of grooves does not change (300 nm for all samples). We believe that this phenomenon is caused by the loss of the adhesion between the film and the substrate due to the nanofillers size. In the case of films with only the nanofillers and without filtration, no protruding ridges, even for 20 μm lines, could be formed, although it is clear that, when the film is underexposed, the line is there, but when DTC is reached, the line loses its adhesion to the substrate. This fact is minimized by the addition of deflocculant, which offered better homogeneity and smaller nanofillers size, resulting in better adhesion to the substrate. The same stands for the filtration of the solution; all the large size nanofillers are retained in the filter, therefore smaller nanofillers are in the film, resulting again in a better adhesion to the substrate.

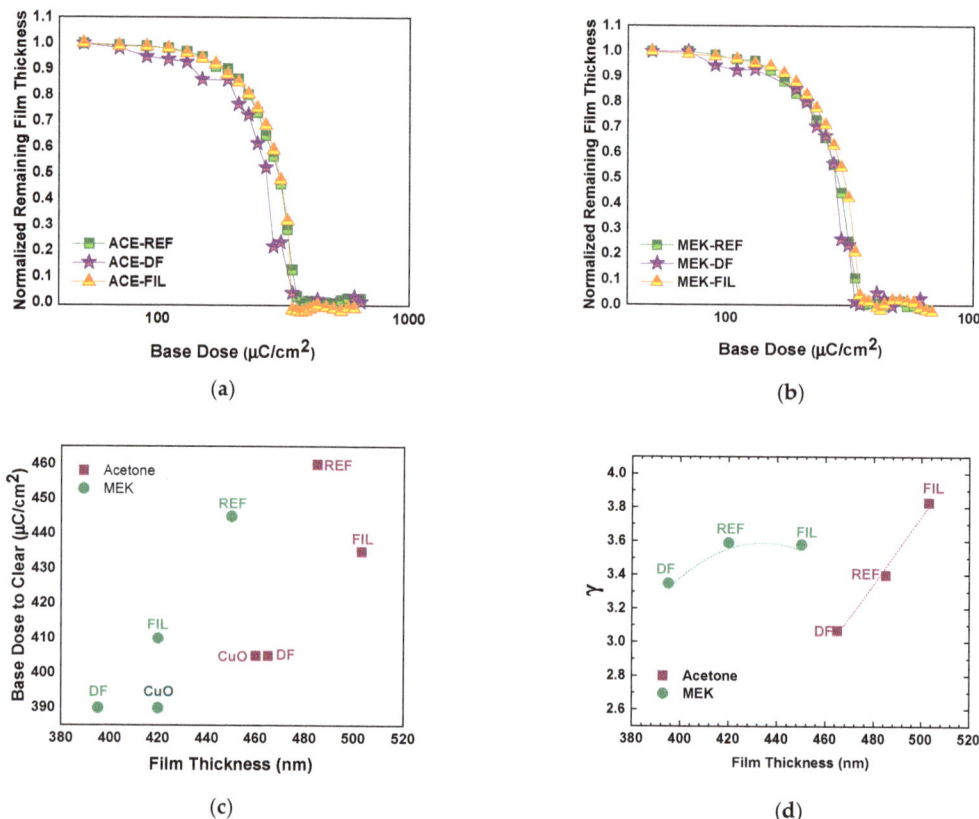

Figure 9. (**a**,**b**) Contrast curves (Normalized Remaining Thickness as a function of the base dose) for the acetone-based and MEK-based nanocomposite films, respectively. (**c**) Sensitivity expressed as minimum base dose-to-clear versus film thickness and (**d**) contrast γ versus film thickness for all nanocomposite films of Phase 2. Lines serve only as a guide to the eye.

A very significant difference between the two solvents was that the MEK-based films were extremely sensitive to the base dose and could only be reliably patterned when the exposure dose was close to DTC (Figure S9). For doses slightly higher than DTC, it was not possible to create protruding ridges smaller than 20 μm, while the grooves, when the line width was smaller than 1 μm, appeared to not be fully exposed. For that reason, MEK was proven to be an unsuitable solvent for this particular methodology. However, acetone resulted in films with increased tolerance to exposure doses (Figure S8), although the resolution limits (300 nm for grooves and 5 μm for protruding ridges) are still rather large for the actual capabilities of EBL. It appears—and in the most dramatic way, this was observed through the reference samples that do not contain any nanofillers—that the addition of the solvent (either MEK or acetone) affects the lithographic properties of PMMA in a profound way, mostly in terms of resolution. Therefore, it is of paramount importance to optimize the quantity of the solvent that may be added to the initial PMMA/PGMEA solution to incorporate the CuO nanofillers.

Summarizing the findings of this study, the first general observation is that, despite its simplicity and "crudeness", the suggested methodology can result in principle in CuO/PMMA nanocomposites that can be patterned with EBL and thus be employed in the future for polymer nanocomposite-based nanodevices. The results showed that nanofiller loadings should not exceed 1% w/v due to the formation of large metal oxide aggregates.

EBL of the CuO/PMMA films showed that the presence of the CuO nanofillers, as well as the addition of solvent (either acetone or MEK), both affect the properties of PMMA by increasing the DTC and its contrast, but putting a significant limit on the resolution. In particular, the results of Phase 2, where the initial PMMA/PGMEA solution concentration was increased from 5% to 6% w/w and the amount of solvent was increased, it was observed that there was a deterioration in the resist resolution (Figures S9 and S10) and a more pronounced aggregation of the nanofillers (Figure 8a), as indicated by the μ-XRF measurements and the calculated CuO areal density (Table 1 vs. Table 2). This indicates that the role of the solvent used to mix the nanofillers into the PMMA/PGMEA and its volume ratio with respect to the initial PMMA/PGMA solution is critical in controlling the properties of the nanocomposite as a resist. Future studies will focus on determining the optimum volume ratio of the added solvent to PMMA/PGMA that will not have a deleterious effect on the resolution of the resist, while minimizing the nanofiller aggregation.

In addition, Phase 2 revealed that MEK is not a suitable choice for this method. Despite the fact that the CuO/PMMA nanocomposites realized using MEK could be patterned by EBL, scrutinizing the resolution patterns revealed that the films were very sensitive to the exposure dose and patterns could only be reliably produced when working very close to DTC, which is a non-desirable feature for any resist. Furthermore, μ-XRF revealed that using MEK leads to the formation of very large aggregates, as indicated by the increase in the CuO areal density by three orders of magnitude (25.5 $\mu g/cm^2$ instead of the anticipated 0.34 $\mu g/cm^2$).

Further, this study indicated that the use of a defloculant offers an improved homogenization and nanofiller dispersion within the polymer matrix and significantly extends the shelf-life of the CuO/PMMA, from 1 day to 1 week. Therefore, it is imperative that the defloculant is used. Again, additional studies are required to optimize the amount of defloculant that would further improve the homogeneity of the nanocomposite, the nanofiller dispersion within the polymer host and possible extension of the shelf-life of the solutions. Filtration ameliorated the films by removing the large aggregates, but unless the initial solutions are more homogenized and with improved nanofiller dispersion, filtration by itself is not enough. Hence, further works studying the combined effect of defloculant use and filtration are required.

4. Conclusions

In this work, the demonstration of an easy and low-cost method to produce e-beam resist materials based on CuO/PMMA nanocomposites was accomplished, showing all the steps from the synthesis of the CuO nanofillers to the development and the evaluation of lithographic performance of the resist. It was established that the suggested method, despite its simplicity, can produce CuO/PMMA nanocomposite EBL resists that can be used in the future for several applications and nanodevices. This study revealed that the most critical parameters are the nature and the volume ratio of solvent with respect to the original PMMA solution, which control both the resolution of the nanocomposite resist and the nanofiller homogeneous dispersion. It was also established that the use of a defloculant is necessary in order to improve the dispersion of the nanofillers, the homogeneity of the nanocomposite solutions and to extend their shelf-life. Filtration of the resist solutions prior to EBL may be beneficial so as to remove the remaining nanofiller aggregates. Finally, apart from the proof-of-concept of the suggested methodology, this work demonstrated that μ-XRF—which was used for the first time according the authors' knowledge in such a context—is a powerful alternative method for the non-destructive and time-efficient characterization of polymer nanocomposite films offering meaningful insights and quantifiable observations. Future works will concentrate on optimizing the key parameters of the methodology, testing of other solvents apart from acetone and MEK, and most importantly extending it to other types of metal oxide nanofillers.

Supplementary Materials: The following are available online at https://www.mdpi.com/2079-4991/11/3/762/s1, Figure S1: Schematic representation of the CuO Nanofiller Synthesis, Figure S2: SEM images of the nanopowders produced with copper (II) acetate monohydrate concentration of (**a**) 30 mM, (**b**) 65 mM and (**c**) 100 mM. Subscripts denote the temperature of synthesis: (1) 70 °C, (2) 80 °C and (3) 90 °C. Scale bar in all images: 100 nm; magnification: ×100,000, Figure S3: (**a**) XRD spectra of the nanopowders compared to the monoclinic phase of CuO (JCPDS pattern no 45-09370) demonstrating the pure monoclinic phase of all samples; (**b**) Crystallite size as function of copper (II) acetate monohydrate concentration for the three synthesis temperatures of 70 °C (black squares), 80 °C (red circles) and 90 °C (blue triangles), Figure S4: Schematics of (**a**) the contrast curve patterns (top view), (**b**) the resolution patterns (top view), and (**c**) the cross-section of the resolution patterns showing the wells and ridges, Figure S5: Photographs of the CuO/PMMA solutions (**a**) 1%CuO/PMMA after production and just prior to use for sample spin-coating, (**b**) 1%CuO/PMMA after 48 h stored in ambient conditions, (**c**) 1%CuO/PMMA with deflocculant after 2 weeks stored in ambient conditions, and (**d**) 1%CuO/PMMA with deflocculant after 1 month stored in ambient conditions, Figure S6: Optical microscope images (×10) of the contrast patterns for the acetone-based CuO/PMMA nanocomposite film, Figure S7: Optical microscope images (×10) of the contrast patterns for the MEK-based CuO/PMMA nanocomposite films. Base doses are indicated on the left-hand side, Figure S8: Optical microscope images (×10) of the resolution patterns for the acetone-based CuO/PMMA nanocomposite films. On the left hand-side the name of the samples is indicated, while the top row indicates the base dose value with proximity effect correction (in $\mu C/cm^2$). Feature size (**a**) L_w = 300 nm, (**b**) L_w = 500 nm, (**c**) L_w = 1 μm, (**d**) L_w = 5 μm, (**e**) L_w = 10 μm and (**f**) L_w = 20 μm, Figure S9: Optical microscope images (×10) of the resolution patterns for the acetone-based CuO/PMMA nanocomposite films. On the left-hand side the name of the samples is indicated, while the top row indicates the base dose value with proximity effect correction (in $\mu C/cm^2$). Feature size (**a**) L_w = 300 nm, (**b**) L_w = 500 nm, (**c**) L_w = 1 μm, (**d**) L_w = 5 μm, (**e**) L_w = 10 μm and (**f**) L_w = 20 μm, Table S1: Summary of CuO nanofiller synthesis conditions per sample.

Author Contributions: Conceptualization, M.C. and E.M.; synthesis of CuO nanofillers, nanocomposite preparation, SEM, sample preparation for XRD, profilometry, G.G.; e-beam lithography and analysis, G.P., G.G., E.M.; XRD measurements and analysis, V.P.; XRF measurements and analysis, A.G.K. and E.M.; writing—original draft preparation, E.M.; writing—review and editing, M.C. and E.M.; supervision, M.C., project administration, E.M. All authors have read and agreed to the published version of the manuscript.

Funding: E.M., V.P., G.P., G.G. and M.C. acknowledge the support of project INNOVATION EL (MIS 5002772), implemented under the "Action for the Strategic Development on the Research and Technological Sector", and project MIS 5002567. Both projects are funded by the Operational Programme "Competitiveness, Entrepreneurship and Innovation" (NSRF 2014–2020) and co-financed by Greece and the European Union (European Regional Development Fund).

Acknowledgments: The authors would like to thank: G.P. Papageorgiou, PhD candidate at the Institute of Nanoscience and Nanotechnology, NCSR "Demokritos" (INN), and the INN clean-room staff for their technical support, Ioannis Raptis, Director of Research at INN for valuable comments on the manuscript and G. Vekinis, Director of Research at INN, for helpful discussions and the procurement of the deflocculant. AK would like to thank Z. Mpari for assistance in the μ-XRF measurements, and S.E. Chatousidou, A. Asvestas and D. Anagnostopoulos for performing hh-XRF measurements with two different analyzers.

Conflicts of Interest: The authors declare no conflict of interest.

References

1. Naskar, A.K.; Keum, J.K.; Boeman, R.G. Polymer matrix nanocomposites for automotive structural components. *Nat. Nanotechnol.* **2016**, *11*, 1026–1030. [CrossRef]
2. Gowri, V.S.; Almeida, L.; De Amorim, M.T.P.; Pacheco, N.C.; Souto, A.P.; Esteves, M.F.; Sanghi, S.K. Functional finishing of polyamide fabrics using ZnO–PMMA nanocomposites. *J. Mater. Sci.* **2010**, *45*, 2427–2435. [CrossRef]
3. Balen, R.; da Costa, W.V.; Andrade, J.D.L.; Piai, J.F.; Muniz, E.C.; Companhoni, M.V.; Nakamura, T.U.; Lima, S.M.; Andrade, L.H.D.C.; Bittencourt, P.R.S.; et al. Structural, thermal, optical properties and cytotoxicity of PMMA/ZnO fibers and films: Potential application in tissue engineering. *Appl. Surf. Sci.* **2016**, *385*, 257–267. [CrossRef]
4. Di Mauro, A.; Cantarella, M.; Nicotra, G.; Pellegrino, G.; Gulino, A.; Brundo, M.V.; Privitera, V.; Impellizzeri, G. Novel synthesis of ZnO/PMMA nanocomposites for photocatalytic applications. *Sci. Rep.* **2017**, *7*, 40895. [CrossRef]

5. Thomas, D.J. Developing nanocomposite 3D printing filaments for enhanced integrated device fabrication. *Int. J. Adv. Manuf. Technol.* **2018**, *95*, 4191–4198. [CrossRef]
6. Gervasio, M.; Lu, K. PMMA–ZnO Hybrid Arrays Using in Situ Polymerization and Imprint Lithography. *J. Phys. Chem. C* **2017**, *121*, 11862–11871. [CrossRef]
7. Mallakpour, S.; Behranvand, V. Nanocomposites based on biosafe nano ZnO and different polymeric matrixes for antibacterial, optical, thermal and mechanical applications. *Eur. Polym. J.* **2016**, *84*, 377–403. [CrossRef]
8. Loste, J.; Lopez-Cuesta, J.-M.; Billon, L.; Garay, H.; Save, M. Transparent polymer nanocomposites: An overview on their synthesis and advanced properties. *Prog. Polym. Sci.* **2019**, *89*, 133–158. [CrossRef]
9. Bueche, A.M. Filler reinforcement of silicone rubber. *J. Polym. Sci.* **1957**, *25*, 139–149. [CrossRef]
10. Fu, S.-Y.; Sun, Z.; Huang, P.; Li, Y.-Q.; Hu, N. Some basic aspects of polymer nanocomposites: A critical review. *Nano Mater. Sci.* **2019**, *1*, 2–30. [CrossRef]
11. Araujo, P. De. Stiffer by Design. *Nat. Mater.* **2007**, *6*, 9–11.
12. Schadler, L.S.; Brinson, L.C.; Sawyer, W.G. Polymer nanocomposites: A small part of the story. *JOM* **2007**, *59*, 53–60. [CrossRef]
13. Kumar, S.K.; Ganesan, V.; Riggleman, R.A. Perspective: Outstanding theoretical questions in polymer-nanoparticle hybrids. *J. Chem. Phys.* **2017**, *147*, 020901. [CrossRef] [PubMed]
14. Krumpfer, J.W.; Schuster, T.; Klapper, M.; Müllen, K. Make it nano-Keep it nano. *Nano Today* **2013**, *8*, 417–438. [CrossRef]
15. Thostenson, E.T.; Li, C.; Chou, T.-W. Nanocomposites in context. *Compos. Sci. Technol.* **2005**, *65*, 491–516. [CrossRef]
16. Hu, Y.; Wu, H.; Gonsalves, K.; Merhari, L. Nanocomposite resists for electron beam nanolithography. *Microelectron. Eng.* **2001**, *56*, 289–294. [CrossRef]
17. Merhari, L.; Gonsalves, K.; Hu, Y.; He, W.; Huang, W.-S.; Angelopoulos, M.; Bruenger, W.; Dzionk, C.; Torkler, M. Nanocomposite resist systems for next generation lithography. *Microelectron. Eng.* **2002**, *63*, 391–403. [CrossRef]
18. Ali, M.A.; Gonsalves, K.E.; Agrawal, A.; Jeyakumar, A.; Henderson, C.L. A new nanocomposite resist for low and high voltage electron beam lithography. *Microelectron. Eng.* **2003**, *70*, 19–29. [CrossRef]
19. Soleimani, E.; Moghaddami, R. Synthesis, characterization and thermal properties of PMMA/CuO polymeric nanocomposites. *J. Mater. Sci. Mater. Electron.* **2018**, *29*, 4842–4854. [CrossRef]
20. Ghashghaee, M.; Fallah, M.; Rabiee, A. Superhydrophobic nanocomposite coatings of poly(methyl methacrylate) and stearic acid grafted CuO nanoparticles with photocatalytic activity. *Prog. Org. Coat.* **2019**, *136*, 105270. [CrossRef]
21. Botsi, S.; Tsamis, C.; Chatzichristidi, M.; Papageorgiou, G.; Makarona, E. Facile and cost-efficient development of PMMA-based nanocomposites with custom-made hydrothermally-synthesized ZnO nanofillers. *Nano-Struct. Nano-Objects* **2019**, *17*, 7–20. [CrossRef]
22. Rabee, B.H.; Al-kareem, B.A. Study of Optical Properties of (PMMA-CuO). *Nanocomposites* **2016**, *5*, 879–883.
23. Hilal, I.H.; Jabbar, R.H.; Muslime, A.H.; Shakir, W.A. Preparation (PMMA/PVA)-copper oxide nanocomposites solar cel. *AIP Conf. Proc.* **2020**, *2290*, 050036. [CrossRef]
24. Katowah, D.F.; AlQarni, S.; Mohammed, G.I.; Al Sheheri, S.Z.; Alam, M.M.; Ismail, S.H.; Asiri, A.M.; Hussein, M.A.; Rahman, M.M. Selective Hg 2+ sensor performance based various carbon-nanofillers into CuO-PMMA nanocomposites. *Polym. Adv. Technol.* **2020**, *31*, 1946–1962. [CrossRef]
25. Wang, D.; Song, C.; Lv, X.; Wang, Y. Design of preparation parameters for commendable photocatalytic properties in CuO nanostructures. *Appl. Phys. A* **2016**, *122*, 1020. [CrossRef]
26. Dang, R.; Jia, X.; Liu, X.; Ma, H.; Gao, H.; Wang, G. Controlled synthesis of hierarchical Cu nanosheets @ CuO nanorods as high-performance anode material for lithium-ion batteries. *Nano Energy* **2017**, *33*, 427–435. [CrossRef]
27. Yu, X.; Marks, T.J.; Facchetti, A. Metal oxides for optoelectronic applications. *Nat. Mater.* **2016**, *15*, 383–396. [CrossRef]
28. Mizuno, J.; Jeem, M.; Takahashi, Y.; Kawamoto, M.; Asakura, K.; Watanabe, S. Light and Shadow Effects in the Submerged Photolytic Synthesis of Micropatterned CuO Nanoflowers and ZnO Nanorods as Optoelectronic Surfaces. *ACS Appl. Nano Mater.* **2020**, *3*, 1783–1791. [CrossRef]
29. Nishino, F.; Jeem, M.; Zhang, L.; Okamoto, K.; Okabe, S.; Watanabe, S. Formation of CuO nano-flowered surfaces via submerged photo-synthesis of crystallites and their antimicrobial activity. *Sci. Rep.* **2017**, *7*, 1–11. [CrossRef]
30. Krcmar, P.; Kuritka, I.; Maslik, J.; Urbanek, P.; Bazant, P.; Machovsky, M.; Suly, P.; Merka, P. Fully Inkjet-Printed CuO Sensor on Flexible Polymer Substrate for Alcohol Vapours and Humidity Sensing at Room Temperature. *Sensors* **2019**, *19*, 3068. [CrossRef]
31. Geng, W.; Ma, Z.; Zhao, Y.; Yang, J.; He, X.; Duan, L.; Li, F.; Hou, H.; Zhang, Q. Morphology-Dependent Gas Sensing Properties of CuO Microstructures Self-Assembled from Nanorods. *Sens. Actuators B Chem.* **2020**, *325*, 128775. [CrossRef]
32. Pargoletti, E.; Cappelletti, G. Breakthroughs in the Design of Novel Carbon-Based Metal Oxides Nanocomposites for VOCs Gas Sensing. *Nanomaterials* **2020**, *10*, 1485. [CrossRef] [PubMed]
33. Ali, F.I.; Mahmoud, S.T.; Awwad, F.; Greish, Y.E.; Abu-Hani, A.F. Low power consumption and fast response H2S gas sensor based on a chitosan-CuO hybrid nanocomposite thin film. *Carbohydr. Polym.* **2020**, *236*, 116064. [CrossRef]
34. Ganga, B.; Santhosh, P. Manipulating aggregation of CuO nanoparticles: Correlation between morphology and optical properties. *J. Alloys Compd.* **2014**, *612*, 456–464. [CrossRef]
35. Mack, C.A.; Legband, D.A.; Jug, S. Data analysis for photolithography. *Microelectron. Eng.* **1999**, *46*, 65–68. [CrossRef]

36. Karydas, A.G.; Czyzycki, M.; Leani, J.J.; Migliori, A.; Osan, J.; Bogovac, M.; Wrobel, P.; Vakula, N.; Padilla-Alvarez, R.; Menk, R.H.; et al. An IAEA multi-technique X-ray spectrometry endstation at Elettra Sincrotrone Trieste: Benchmarking results and interdisciplinary applications. *J. Synchrotron Radiat.* **2018**, *25*, 189–203. [CrossRef]
37. Solé, V.; Papillon, E.; Cotte, M.; Walter, P.; Susini, J. A multiplatform code for the analysis of energy-dispersive X-ray fluorescence spectra. *Spectrochim. Acta Part B At. Spectrosc.* **2007**, *62*, 63–68. [CrossRef]
38. Singh, B.P.; Bhattacharjee, S.; Besra, L.; Sengupta, D.K. Evaluation of dispersibility of aqueous alumina suspension in presence of Darvan C. *Ceram. Int.* **2004**, *30*, 939–946. [CrossRef]
39. Singh, B.P.; Nayak, S.; Samal, S.; Bhattacharjee, S.; Besra, L. The role of poly(methacrylic acid) conformation on dispersion behavior of nano TiO2 powder. *Appl. Surf. Sci.* **2012**, *258*, 3524–3531. [CrossRef]
40. Bukhari, S.Z.A.; Ha, J.-H.; Lee, J.; Song, I.-H. Viscosity Study to Optimize a Slurry of Alumina Mixed with Hollow Microspheres. *J. Korean Ceram. Soc.* **2015**, *52*, 403–409. [CrossRef]
41. Gaikwad, A.M.; Khan, Y.; Ostfeld, A.E.; Pandya, S.; Abraham, S.; Arias, A.C. Identifying orthogonal solvents for solution processed organic transistors. *Org. Electron.* **2016**, *30*, 18–29. [CrossRef]

Review

Review and Mechanism of the Thickness Effect of Solid Dielectrics

Liang Zhao [1] and Chun Liang Liu [2,*]

1. Science and Technology on High Power Microwave Laboratory, Northwest Institute of Nuclear Technology, P.O. Box 69 Branch 13, Xi'an 710024, China; zhaoliang@stu.xjtu.edu.cn
2. Key Laboratory of Physical Electronics and Devices of Ministry of Education, Xi'an Jiaotong University, No. 28 West Xianning Rd., Xi'an 710049, China
* Correspondence: chlliu@mail.xjtu.edu.cn

Received: 3 November 2020; Accepted: 8 December 2020; Published: 10 December 2020

Abstract: The thickness effect of solid dielectrics means the relation between the electric breakdown strength (E_{BD}) and the dielectric thickness (d). By reviewing different types of expressions of E_{BD} on d, it is found that the minus power relation ($E_{BD} = E_1 d^{-a}$) is supported by plenty of experimental results. The physical mechanism responsible for the minus power relation of the thickness effect is reviewed and improved. In addition, it is found that the physical meaning of the power exponent a is approximately the relative standard error of the E_{BD} distributions in perspective of the Weibull distribution. In the end, the factors influencing the power exponent a are discussed.

Keywords: thickness effect; solid insulation dielectrics; breakdown strength

1. Introduction

Solid dielectrics are widely used in high-voltage (HV) devices and pulsed power systems [1–6]. The breakdown characteristics of solid dielectrics have been researched for nearly 100 years. The classical solid dielectric breakdown theory includes intrinsic breakdown, avalanche breakdown, thermal breakdown, and electro-mechanical breakdown, which were established by A. von. Hippel [7–14], H. Frohlich [15–21], and F. Seitz [22–28] and later refined by G. C. Garton [29], S. Whitehead [30,31], and J. J. O'Dwyer [32–34]. The classical breakdown theory mainly focuses on questions such as the difference between intrinsic breakdown and avalanche breakdown, the relation between electric breakdown strength (E_{BD}) and dielectric thickness (d), the dependence of E_{BD} on temperature (T), and the tendency of E_{BD} in applied waveforms, etc. Table 1 summarizes these types of breakdown.

From this table, it can be seen that a common question is the dependence of E_{BD} on dielectric thickness d. If E_{BD} is independent of d, the breakdown may be classified as the intrinsic type; if not, the breakdown may be classified to other types. Thus, research on the thickness effect of solid dielectrics is important in breakdown theory. In practice, research on the thickness effect is also important for practical insulation design, since E_{BD} is directly related to the size (namely, thickness) and the lifetime of solid insulation structures.

In view of these considerations, understanding the mechanism of the thickness effect and knowing the specific expression of E_{BD} on d are meaningful. However, research on these topics is not ideal and systematic, even thought a lot of experiments have been reported.

Table 1. Summary of classical breakdown theory.

Breakdown Type	General Mechanism	Basic Characteristics
Intrinsic breakdown	"Electron instability"	1. E_{BD} is independent of d. 2. The breakdown time is in the nanosecond time scale. 3. The breakdown happens in a low temperature range.
Avalanche breakdown	Electron impact and ionization	1. E_{BD} is dependent on d and the electrode configuration. 2. The breakdown time is in the nanosecond time scale. 3. The breakdown happens in a low temperature range.
Thermal breakdown	"Heat instability"	1. The breakdown time is longer (in the microsecond time scale or longer). 2. E_{BD} is related to sample and electrode waveform. 3. The breakdown happens in a high temperature range.
Electro-mechanical breakdown	Electro-mechanical force	1. It is common for plastics and crystals. 2. It happens easily when defects exist in dielectrics.

The main questions can be classified into the following aspects:

(1) The relation between E_{BD} on d is not unified. For example, at least four expressions related to the effect of E_{BD} on d have been reported.

(2) The mechanism responsible for the thickness effect is not well understood. This is probably why different expressions exist.

(3) The application condition for some types of E_{BD}-d expressions is not clear.

Keeping in mind these questions above, a review and theoretical analysis are presented in this paper. In Sections 2 and 3, different expressions regarding the relation of E_{BD} with d are reviewed and compared, aiming to find the most appropriate one. In Section 4, the physical model for the minus power relation of the thickness effect, which is considered as the most appropriate one, is reviewed and improved. In Section 5, the thickness effect is interpreted in light of the statistics distribution. In Section 6, the power exponent in the minus power relation of the thickness effect is discussed. The last question is discussed in the conclusions of this paper.

Before starting the review, it is noted that "solid dielectrics" in this paper mainly denote the solid insulation dielectrics, which cannot be easily transformed and can play both the role of insulation and support in HV devices, such as polymethyl methacrylate (PMMA), polyethylene (PE), nylon, Al_2O_3, MgO, TiO_2. Solid dielectrics of elastomers such as polymerized styrene butadiene rubber, cis-polybutadiene, and polyisoprene rubber are not discussed in this paper [35]. This is not only because these types of dielectrics have an unstable thickness but also because there are several breakdown mechanisms involved in the failure process aside from electric breakdown [36], such as thermal breakdown and electro-mechanical breakdown.

2. Review on Reported Relations about E_{BD} on d

At least four types of relations between E_{BD} and d have been reported in the literature—i.e., constant relation, reciprocal-single-logarithm relation, minus-single-logarithm relation, and double logarithm relation. Each relation has its own experimental supports.

2.1. Constant Relation

The constant relation means that E_{BD} is equal to a constant C, which is independent on d—i.e.,

$$E_{BD}(d) = C. \qquad (1)$$

Two groups of experimental results support this type of relation. The first group comes from W. J. Oakes [6] in 1948, who tested the breakdown voltage U_{BD} of PE samples ranging from 0 to 0.2 mm under dc conditions. He found that U_{BD} was linearly dependent on d, which means that E_{BD} is a constant. The second group of experimental results comes from J. Vermeer [37] in 1954, who tested the E_{BD} of Pyrex glass and found that, when the temperature is −50 °C or the voltage rise time t_r is equal to 10^{-5} s, E_{BD} would be independent of d. These two groups of experimental results are re-analyzed and re-plotted in Figure 1.

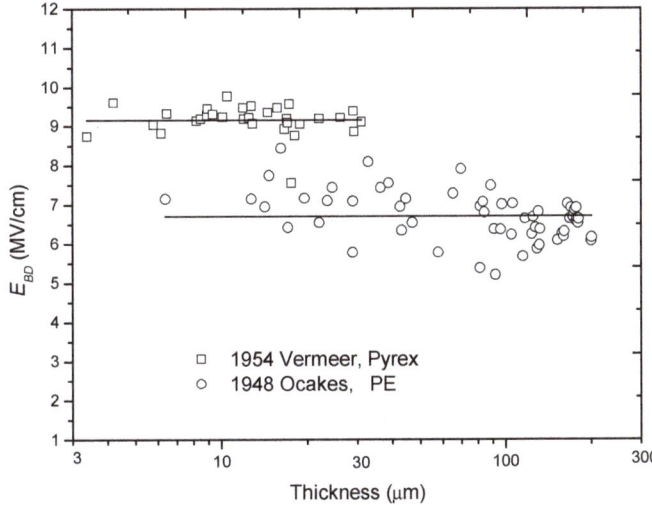

Figure 1. Experimental results supporting the constant relation of the dielectric field E_{BD} with thickness d.

The constant relation of E_{BD} on d may reflect some intrinsic breakdown characteristics of dielectrics.

2.2. Reciprocal-Single-Logarithm Relation

The reciprocal-single-logarithm relation means that $1/E_{BD}$ is linear to $\lg d$—i.e.,

$$1/E_{BD}(d) = A\lg d - B. \qquad (2)$$

where A and B are constants. This relation is based on the "40-generation-electron theory", which was put forward by F. Seitz in 1949 [38]. The deduction process for Equation (2) is presented concisely as follows: a seed electron can prime an electron avalanche after 40-time impact and ionization—i.e.,

$$\alpha d = 40, \qquad (3)$$

where α is the ionization coefficient, which can be written as follows:

$$\alpha \propto \exp\left(-\frac{E_H}{E}\right), \qquad (4)$$

where E_H is a parameter with the same unit of electric field and is dependent on E. Inserting Equation (4) into Equation (3) gives the following:

$$d \exp\left(-\frac{E_H}{E}\right) = d_0(E), \tag{5}$$

where $d_0(E)$ means a unit thickness dependent on the applied field E. Based on Equation (5), E is solved and defined as E_{BD}, which is as follows:

$$E_{BD} = \frac{E_H}{\ln[d/d_0(E_{BD})]}. \tag{6}$$

As to Equation (6), both sides have E_{BD} and this equation cannot be solved. Even still, it is widely accepted that E_H and $d_0(E)$ can be considered as constants. Based on this, Equation (6) can be further changed to:

$$\frac{1}{E_{BD}} = \frac{1}{E_H} \ln d - \frac{\ln d_0}{E_H}. \tag{7}$$

By comparing (7) and (2), one can see that $A = 1/E_H$ and $B = 1/E_H \ln d_0$.

Additionally, there are two groups of experimental results supporting this type of relation. The first group comes from A. W. E. Austen in 1940 [30], who tested the E_{BD} of clean ruby muscovite mica dependent on a thickness in a range of 200–600 nm under dc conditions. He found that:

$$E_{BD} = \frac{54}{\ln(d/d_0)} (\text{MV/cm}), \tag{8}$$

where $d_0 = 5$ nm. The second group comes from J. J. O'Dwyer in 1967 [33], who thoroughly analyzed the electron-impact-ionization model of Seitz and summarized the relevant experimental results in history to support the E_{BD}-d relation in (2). Figure 2 replots the two groups of experimental results with $1/E_{BD}$ dependent on lg$d$.

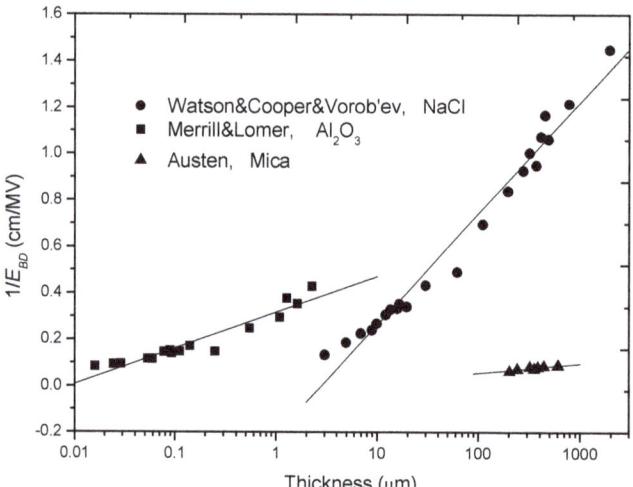

Figure 2. Experimental results supporting the reciprocal-single-logarithm relation of E_{BD} with d.

2.3. Minus-Single-Logarithm Relation

The minus-single-logarithm relation means that E_{BD} is linear to lg$d$ with a minus slope rate—i.e.,

$$E_{BD}(d) = D - F \lg d, \tag{9}$$

where D and F are constants. In 1963, R. Cooper tested the E_{BD} of PE with a thickness ranging from 25 to 460 μm on a millisecond time scale [39]. After fitting, Cooper gave the following relation:

$$E_{BD}(d) = 12.8 - 3 \lg d \, (\text{MV/in}), \tag{10}$$

where d is in inches. It is noted that this result was published in *Nature* but there was no theoretical basis for Equation (9) or Equation (10). Figure 3 replots the experimental results with the units of MV/cm and cm.

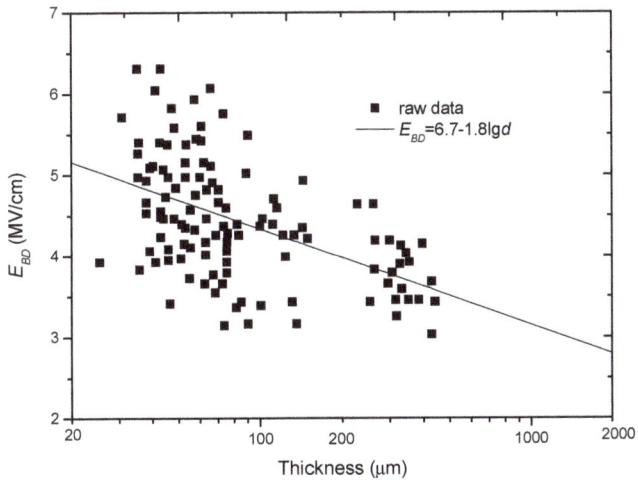

Figure 3. Experimental results to support the minus-single-logarithm relation of E_{BD} with d.

2.4. Double-Logarithm Relation

The double-logarithm relation means that $\lg E_{BD}$ is linear to $\lg d$—i.e.,

$$\lg E_{BD}(d) = G - a \lg d. \tag{11}$$

where G and a are constants. The expression in Equation (10) is also called the minus power relation since it can be transformed into:

$$E_{BD}(d) = E_1 d^{-a}, \tag{12}$$

where $E_1 = 10^G$. In 1964, F. Forlani put forward a model for the thickness effect by taking into account the electron injection from cathode and the electron avalanche in dielectrics together [40,41]. After a series deduction, he found that:

$$E_{BD} \approx \frac{C_\phi}{d^a}. \tag{13}$$

According to Forlani, C_φ and a are constants and a ranges from 1/4 to 1/2. By re-analyzing the experimental data from R. C. Merrill on Al_2O_3, he verified this model. Here, we also replot this group of data in a log-log coordinate system, as shown in Figure 4.

As a summary of this section, Table 2 lists the expressions, mechanisms, and researchers for the formulae of the thickness effect in the literature.

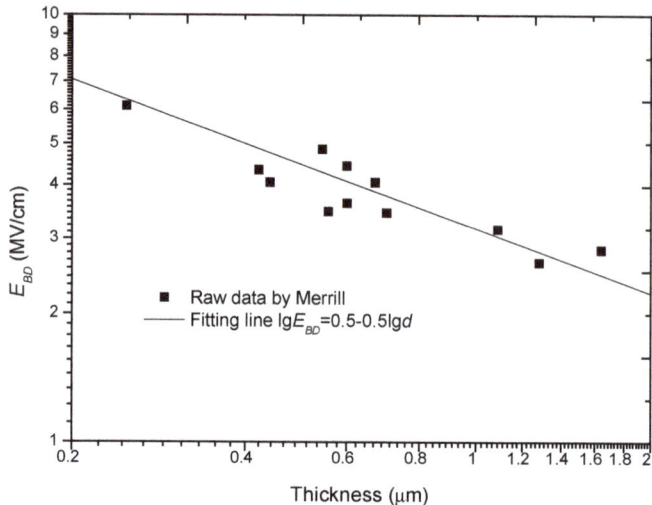

Figure 4. Experimental results to support the double-logarithm relation of E_{BD} with d.

Table 2. Four typical relations for the thickness effect of solid dielectrics.

Typical Relation	Mathematical Expression	Mechanism	Researcher/Year
Constant relation	$E_{BD}(d) = C$	Intrinsic breakdown	Oakes/1948 [6] Vermeer/1954 [37]
Reciprocal-single-logarithm relation	$1/E_{BD}(d) = A\lg d - B$	Avalanche breakdown	Austen/1940 [30] O'Dwyer/1967 [33]
Minus-single-logarithm relation	$E_{BD}(d) = D - F\lg d$.	/	Cooper/1963 [39]
Double-logarithm relation or minus power relation	$\lg E_{BD}(d) = G - a\lg d$.	Electron injection and avalanche	Forlani/1964 [40] Merrill/1963 [42]

3. Comparison of Different E_{BD}-d Relations

Now a question comes: which one is more appropriate to characterize the thickness effect of solid dielectrics? In this section, these four types of E_{BD}-d relations are compared in order to find the most appropriate one.

3.1. More Results on $\lg E_{BD}$-$\lg d$ Relation

When reviewing the thickness effect of solid dielectrics, it is found that plenty of experimental results have been reported on this topic. Among these groups of data, some were already fitted with the $\lg E_{BD}$-$\lg d$ relation and were given the power exponent, such as those reported by J. H. Mason [43–45], G. Yilmaz [46,47], and A. Singh [48,49]; other groups of data were re-analyzed in order to obtain the minus power exponent, such as the data reported by Y. Yang [50], K. Yoshino [51], K. Theodosiou [52], and G. Chen [53]. As a summary, Table 3 lists the researcher, test object, thickness range, value of a, and feature of each group of experiments. Based on this table, the distribution of a in a wide range of thickness is plotted, as shown in Figure 5.

Table 3. Summary of the experimental results for the relation of $E_{BD} = E_1 d^{-a}$ for the thickness effect.

Year/Researcher	Test Object and Condition	Thickness Range	Value of a	Comments/Feature
1948/Oakes	PE, ac	22 μm–350 μm	$a = 0.47$ [6]	
1961/Cooper	NaCl	236 μm–544 μm	$a = 0.33$ [55]	
1963/Vorob'ev	NaCl	3 μm–20 μm	$a = 0.60$ [55]	
1965/Watson	NaCl	24 μm–2000 μm	$a = 0.33$ [55]	In mm range.
1950/Lomer	Al_2O_3	13 nm–0.154 μm	$a = 0.20$ [55]	
1963/Merrill	Al_2O_3	0.25 μm–2.5 μm	$a = 0.50$ [42]	In Å range.
1968/Nicol	Al_2O_3	0.15 nm–60 nm	$a = 0.20$ [55]	
1955/Mason	PE, 1/25 μs pulse	0.1 mm–6.5 mm	$a = 0.66$ [56,57]	In mm range.
1971/Agarwal	Barium stearate, dc	2.5 nm–25 nm	$a = 1.0$ [54]	a is the largest.
1971/Agarwal	Barium stearate, dc	25 nm–200 nm	$a = 0.59$ [54]	
1979/Yoshino	Hexatriacontane, 6 μs	14 μm–100 μm	$\alpha = 0.66$ [51]	
1982/Singh	MgO, ac	4 nm–20 nm	$a = 0.23$ [48]	In nm range.
1983/Singh	La_2O_3, ac	4 nm–40 nm	$a = 0.66$ [48]	
1983/Baguji	TiO_2, ac	40 nm–200 nm	$a = 0.55$ [49]	
1991/Mason	PP, ac, φ63.5 mm	8 μm–76 μm	$a = 0.24$ [44]	Reflecting the factor of electrode on E_{BD}.
1991/Mason	PP, ac, φ12.5 mm	8 μm–76 μm	$a = 0.33$ [44]	
1991/Mason	PP, ac, φ10 mm v.s. φ10 mm	100 μm–500 μm	$a = 0.5$ [44]	
1991/Mason	PVC, dc, ε_r of liquid is 9.	40 μm–500 μm	$a = 0.33, 0.38$ [44]	Reflecting factor of ambient liquid.
1991/Mason	PVC, dc, ε_r of liquid is 5.	40 μm–500 μm	$a = 0.66, 0.70$ [44]	
1992/Helgee	PI, ac	13 μm–270 μm	$a = 0.39$ [58]	
1992/Helgee	PEI, ac	13 μm–270 μm	$a = 0.44$ [58]	
1992/Helgee	PET, ac	13 μm–270 μm	$a = 0.47$ [58]	
1992/Helgee	PEEK, ac	13 μm–270 μm	$a = 0.48$ [58]	
1992/Helgee	PES, ac	13 μm–270 μm	$a = 0.51$ [58]	
1996/Yilmaz	PES, ac	12 μm–200 μm	$a = 0.26$–0.32 [59]	
1997/Yilmaz	PES, ac (0 °C)	100 μm–200 μm	$a = 0.28$ [47]	Focusing on the factor of temperature.
1997/Yilmaz	PES, ac (80 °C)	100 μm–200 μm	$a = 0.30$ [47]	
1997/Yilmaz	PES, ac (120°C)	100 μm–200 μm	$a = 0.32$ [47]	
2003/Yang	TiO_2, dc	100 μm–300 μm	$a = 0.97$ [50]	a is the largest.
2004/Theodosiou	PET, dc	25 μm–350 μm	$a = 0.50$ [52]	
2010/Diaham	PI, dc	1.4 μm–6.7 μm	$a = 0.16$–0.25 [60]	
2011/Zhao	PMMA, PE, Nylon, PTFE *, ns pulse	0.5 mm–3.2 mm	$a = 0.125$ [61]	In nanosecond pulse
2012/Chen	PE, dc	25 μm–250 μm	$a = 0.022$ [53]	a is the smallest.
2013/Neusel	Al_2O_3, TiO_2, $BaTiO$, $SrTiO_3$	2 μm–2 mm	$a = 0.5$ [62]	Plenty of dielectrics were tested.
2013/Neusel	PMMA, PS **, PVC ***, PE	2 μm–2 mm	$a = 0.5$ [62]	

* PTFE: Poly tetra fluoroethylene; ** PS: Polystyrene; *** PVC: Polystyrene.

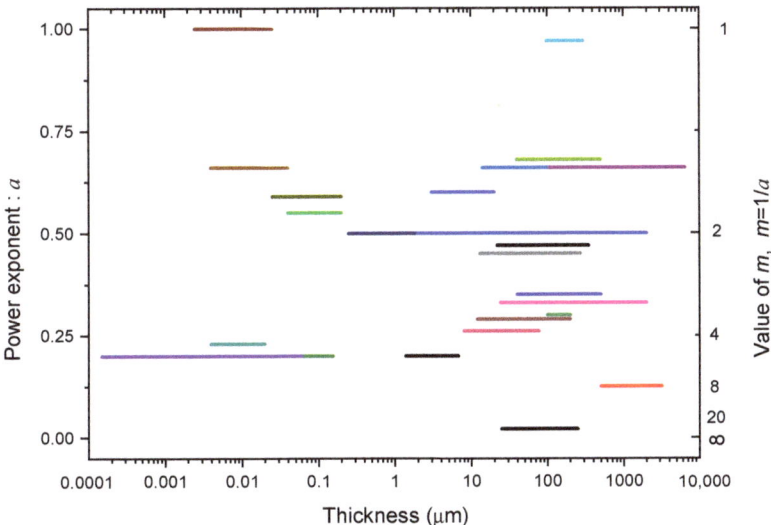

Figure 5. The distribution of a or $m(=1/a)$ in a wide thickness range based on the data summarized in Table 3.

With Table 3 and Figure 5 together, some basic conclusions can be drawn:

(1) The minus power relation for the thickness effect of solid dielectrics holds true in a wide thickness range from several Å s to several millimeters, even though the power exponent a is different.

(2) The value of a in the minus power relation ranges from 0 to 1. The largest value is about 1, see the lines of 1971/Agarwal [54] and 2003/Yang [50]. The smallest value is 0.022; see the line of 2012/Chen. If other factors are neglected, a is averaged at about 0.5.

It is worth mentioning that the distribution of a in thickness seems "random". This will be discussed especially in Section 7.

According to Section 3.1, it can be seen that the number of experimental groups needed to support the minus power relation is far more than those needed to support three other types of thickness–effect relations. Thus, it is necessary to make a comparison between the minus power relation and other three types of relations on the thickness effect.

3.2. Comparison between Minus Power Relation and Other Three Types of Relations

3.2.1. Comparison with the Reciprocal-Single-Logarithm Relation

Figure 6 directly shows the fitting results on the raw experimental data from Austen [30] with the double-logarithm relation and the reciprocal-single-logarithm relation. From Figure 6a, it seems that the two types of fitting have no difference due to the distribution and error bar of the raw data. In order to show the fitting results clearly, the error bar is removed and only the average data are plotted in a log-log coordinate system, as shown in Figure 6b. From this figure, it is seen that the reciprocal-single-logarithm relation gives a smaller E_{BD} in the middle thickness range, whereas it presents a larger E_{BD} in the lower and the higher thickness ranges.

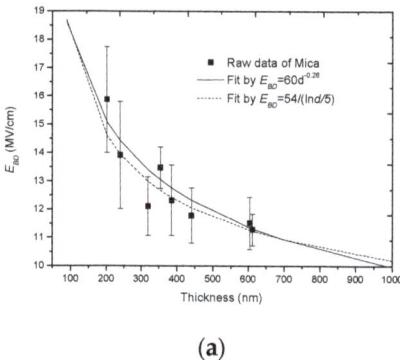
(a)

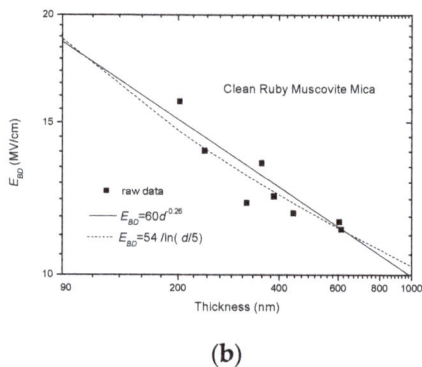
(b)

Figure 6. Comparison between the minus power relation and the reciprocal-single-logarithm relation with raw data from Austen [30]. (**a**) Two types of fitting in a linear coordinate system, (**b**) two types of fitting in a log-log coordinate system.

Aside from the experimental data from Austen, the data from O'Dwyer are also compared, as shown in Figure 7a,b. From these two figures, it can be seen that the minus power relation gives a better fit in all the data ranges, whereas the reciprocal-single-logarithm relation can only cover parts of the experimental data range. In addition, there is a pre-condition to use the reciprocal-single-logarithm relation. Because $\log d/d_0$ should be positive, d should be larger than d_0, or else a minus E_{BD} would be result from this.

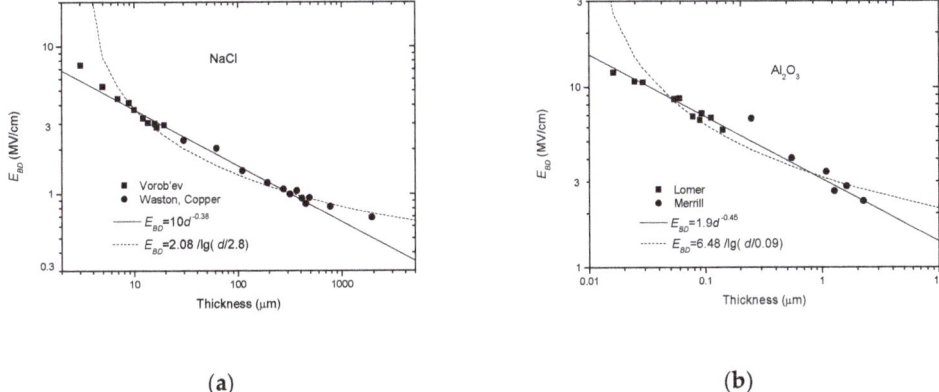

(a)   (b)

Figure 7. Comparison between the minus power relation and the reciprocal-single-logarithm relation with experimental data from O'Dwyer in a log-log coordinate system [33]. (a) NaCl, (b) Al$_2$O$_3$.

Due to these considerations, it is believed that the minus power relation is preferable to the reciprocal-single-logarithm relation to describe the thickness effect of solid dielectrics.

3.2.2. Comparison with the Minus-Single-Logarithm Relation

Similarly, the raw experimental data from R. Cooper are used and fitted with the minus power relation and the minus-single-logarithm relation, respectively, which are shown in Figure 8. From this figure, it can be seen that both the two types of relations can pass the main data range. However, the E_{BD} fitted in the lower and the higher thickness ranges by the minus-single-logarithm relation is lower than that fitted by the minus power relation, and the deviation becomes greater as d deviates from the average thickness of 100 µm; whereas, the minus power relation can be applied in a wide thickness range. In view of this, the minus power relation is also believed to be preferable to the minus-single-logarithm relation. It is noted that the dispersion of data is very high. This is probably due to the two types of waveforms applied (1/8000 µs and 1/120 µs), the different grades of polythene (sample material), the faults in the sample, and the systematic errors in measurement [39].

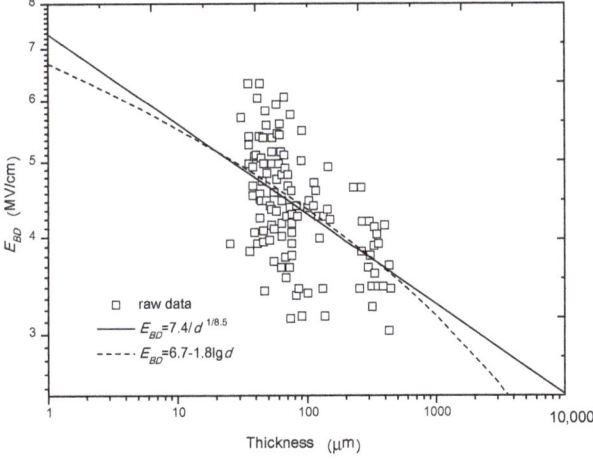

Figure 8. Comparison between the minus power relation and the minus-single-logarithm relation with the experimental data from Cooper in a log-log coordinate system.

3.2.3. Comparison with the Constant Relation

It is believed that the constant relation for the thickness effect is just an extreme case of the minus power relation when a is close to 0. As to Equation (13), if $a = 0$, $E_{BD}(d) = C_\varphi$, which is a constant.

This transition from the minus power relation to the constant relation can be verified by the experimental results by G. Chen in Table 3 [53], since the value of a is 0.022, which means a much weaker thickness effect.

As a conclusion for this section, the minus power relation is believed to be preferable to the other three types of relations to describe the thickness effect on E_{BD}.

4. Mechanism for the Minus Power Relation

If the minus power relation is the most appropriate relation to describe the thickness effect, what is the potential mechanism? As mentioned previously, F. Forlani put forward a physical model to explain the thickness effect and deduced the minus power relation with a theoretical power exponent from 1/4 to 1/2 in 1964. However, the practical range of a is from 0 to 1. This deviation needs to be analyzed and discussed. Before the analysis, the physical model by Forlani is reviewed first.

4.1. Review on Model Suggested by F. Forlani

In Forlani's model, the electrodes and the dielectric were considered together for the occurrence of a breakdown [40,41]. The basic starting point of this model can be written as follows:

$$j(d) = j_i P \exp(\alpha d), \tag{14}$$

where $j(d)$ represents the current density when the seed electrons leave the cathode with a distance of d; j_i represents the current density near the cathode; P denotes the probability for electrons to change from the stable state to an unstable state, which can also be considered as the probability of an avalanche forming; α is the ionization coefficient, which means that α times of impacts to the atoms can take place when an electron moves a distance of 1 cm along the inverse field direction in dielectrics; $\exp(\alpha d)$ represents the increasing times for one seed electron moving along a distance of d. The physical meaning of Equation (14) can be explained as follows: j_i electrons of an initial electron number of j_0 are emitted from the cathode to a dielectric due to the weakness of the potential barrier of the cathode; after moving a distance of d in the dielectric, the electron number becomes $j_i\exp(\alpha d)$ due to the impact ionization and multiplication of $\exp(\alpha d)$. The final electron number in the avalanche head is $j_i\exp(\alpha d)P$ due to the avalanche formation probability P. The schematics of this model are expressed in Figure 9. Namely, there are basically two steps for a breakdown to take place: firstly, the electron injection process of cathode, which is written as Step 1; secondly, the avalanche process in dielectrics, which is written as Step 2. The second step combines two sub-processes together—the electron multiplication process and the avalanche formation process—which are written as Step 2a and Step 2b. It is noted that Step 2a and Step 2b in practice cannot be divided. Here, this is just for the convenience of the deduction of E_{BD} in Section 4.2.

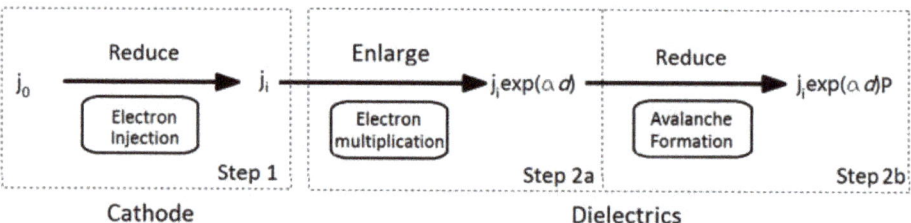

Figure 9. Schematics of the breakdown model by F. Forlani.

If the current density (or electron number) increases to a critical level j_{BD}, which can evaporate or erode the local dielectric, Forlani believed that breakdown occurs—that is:

$$E_{BD} = E\big|_{j(d)=j_{BD}}. \tag{15}$$

Taking into account Equation (15) for Equation (14) and making a logarithm transformation for Equation (14) gives:

$$\ln j_i + \alpha d + \ln P = \ln j_{BD}. \tag{16}$$

Now, each part in the left side of Equation (16) is specially analyzed.

(1) As to j_i, it represents Step1 and is related to the way of electron injection. Forlani takes into account two typical means of electron injection.

The first way is field-induced emission. When the applied field is higher than 1 MV/cm, the field-induced emission is mainly the tunnel effect, which can be expressed as follows:

$$j_i = AE^2 \exp\left(-\frac{B}{E}\theta(y)\right), \tag{17}$$

or

$$\ln j_i = \ln(AE^2) - \frac{B\theta(y)}{E}, \tag{18}$$

where $\theta(y)$ is a modified factor which is related to temperature T. If the T is fixed, $\theta(y)$ can be considered to be a constant.

The second way is the field-assisted thermal emission. Based on the Schottky effect, j_i can be expressed as follows:

$$j_i = j_0 \exp\left(-\frac{\Phi}{kT} + \frac{0.44}{T}E^{1/2}\right), \tag{19}$$

or

$$\ln j_i = \ln j_0 - \frac{\Phi}{k_B T} + \frac{0.44}{T}E^{1/2}, \tag{20}$$

where Φ is the potential barrier of the cathode and k_B is Boltzmann's constant.

(2) As to αd, it represents Step 2a and Forlani believed that α was related to the applied field E; this can be written as follows:

$$\alpha d = \frac{eEd}{\Delta I}, \tag{21}$$

where e is the absolute electron charge and ΔI is the ionization energy.

(3) As to avalanche formation probability P, it represents Step 2b. Forlani solved the wave number equation:

$$\frac{1}{k}\overline{\left(\frac{dk_x}{dt}\right)_{sc}\left(\frac{d\Delta I}{dt}\right)_{sc}} = \left(\frac{eE}{m*}\right)^2, \tag{22}$$

where k is the wave vector, k_x is the wave vector at a distance of x, and $m*$ is the effective mass of the electron. He obtained the curve of P as a function of E, which is replotted in Figure 10. In this figure, E_H represents the breakdown field deduced from the Frohlich low-energy criterion.

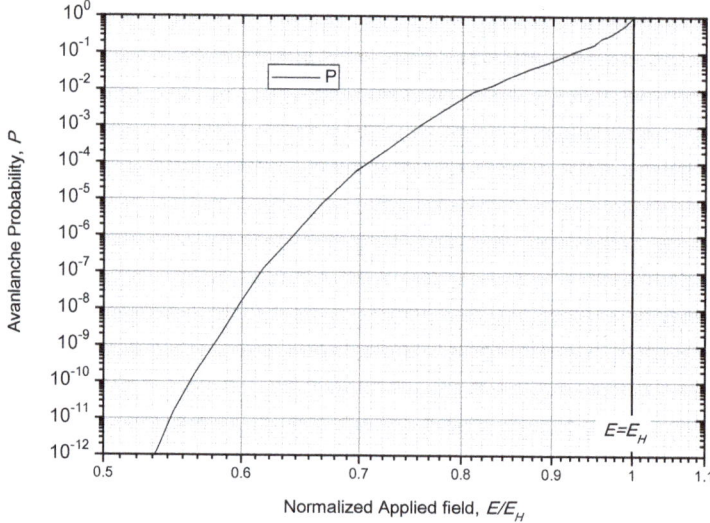

Figure 10. The dependency of avalanche formation probability P on the normalized applied field E/E_H [40,41].

4.2. Solution and Improvements for Forlani' Model

Forlani solved Equation (16) in three cases. First, when the dielectric thickness d was small and the electron injection was mainly via the tunnel effect. According to Forlani, d should be far smaller than the electron recombination length x_0. In this case, Equation (18) can be inserted into Equation (16), i.e.,

$$\frac{eEd}{\Delta I} - \frac{B\theta(y)}{E} + \ln(AE^2) + \ln P = \ln j_{BD}. \qquad (23)$$

Since the applied field is large, P is close to 1 and $\ln P = 0$. In addition, $\ln(AE^2)$ and $\ln j_{BD}$ are smaller parts compared with the first two parts in the left side of Equation (23). Thus, they can be ignored. Taking into account all of these, Equation (23) can be simplified to be:

$$\frac{eEd}{\Delta I} - \frac{B\theta(y)}{E} \approx 0. \qquad (24)$$

Thus, the solution for Equation (24) is:

$$E_{BD} \approx E_{TN} d^{-\frac{1}{2}} \quad \text{where} \quad E_{TN} = \left(\frac{B\theta(y)}{e}\right)^{\frac{1}{2}}. \qquad (25)$$

In Equation (25), E_{TN} is a constant which is related to the tunnel effect.

Second, when the dielectric thickness is small and the electron injection is mainly via the Schottky effect, Equation (20) can be inserted into Equation (16), which gives:

$$\frac{eEd}{\Delta I} - \frac{\Phi}{kT} + \frac{0.44}{T} E^{1/2} + \ln P + \ln j_0 = \ln j_{BD}. \qquad (26)$$

Similarly, since E is large, $\ln P$ is close to 0. In addition, Forlani believed that, by reasonable estimation, only the first two parts in the left side of Equation (25) are predominant—i.e.,

$$\frac{eEd}{\Delta I} - \frac{\Phi}{kT} \approx 0. \qquad (27)$$

Solving (27) gives:

$$E_{BD} \approx E_{ST}d^{-1} \quad \text{where} \quad E_{ST} = \frac{\Phi \Delta I}{k_B T e}, \tag{28}$$

where E_{ST} is a constant which is related to the Schottky effect.

Third, when the dielectric thickness is large and the electron injection from cathode can be neglected, only the electron multiplication part ad and the avalanche formation probability $\ln P$ in Equation (16) need to be taken into account. Additionally, by neglecting $\ln j_{BD}$, one can obtain:

$$\frac{eEd}{\Delta I} + \ln P \approx 0. \tag{29}$$

In order to solve Equation (29), the specific effect of $\ln P$ on E should be known in advance. Unfortunately, Forlani did not give the theoretical relation of P with E, only presenting the curve of $(-\ln P)$ and E, as shown in Figure 10. In addition, Forlani believed that $-\ln P$ is proportional to $1/E^3$. Those are all the clues about P and E supplied by Forlani.

By analyzing these clues, it is found that $\lg(-\ln P)$ should be linear to $\lg E$ with a slope rate of -3. Thus, we get the raw data in Figure 10 and replot the data of $(-\ln P, E)$ in a log-log coordinate system, as shown in Figure 11. From this figure, it can be seen that $\lg(-\ln P)$ is really linear to $\lg E$ with a slope rate of -3, but this relation only holds true within $E/E_H < 0.8$. In order to present an accurate fit, we follow the method suggested by Forlani and consider each segment of the curve in Figure 11 to be a line and fit these segmental lines with a different minus power relation.

Here, a five-segment approximate curve is given, which is:

$$-\ln P = \begin{cases} \frac{K_1}{E^{3.0}} & (0.45E_H \leq E \leq 0.8E_H) \\ \frac{K_2}{E^{5.1}} & (0.80E_H < E \leq 0.9E_H) \\ \frac{K_3}{E^{8.7}} & (0.9E_H < E \leq 0.95E_H) \\ \frac{K_4}{E^{26}} & (0.95E_H < E \leq 0.99E_H) \\ \frac{K_5}{E^{213}} & (0.99E_H < E \leq 1.0E_H) \end{cases}. \tag{30}$$

As to Equation (30), two points need to be clarified: (1) the more segments are set, the more accurate the approximate curve is; (2) the closer E is to E_H, the larger the power exponent is.

As to each segment of the fitting curve, the following relation can be obtained:

$$-\ln P = \frac{K_i}{E^{\omega_i}} \quad p_i < E/E_H \leq p_{i+1}. \tag{31}$$

In this range of E, we solve Equation (29) and obtain:

$$E_{BD} \approx E_i d^{-\frac{1}{\omega_i+1}} \quad \text{where} \quad E_i = \left(\frac{\Delta I K_i}{e}\right)^{\frac{1}{\omega_i+1}}. \tag{32}$$

If m is equal to $\omega_i + 1$, Equation (32) changes to:

$$E_{BD} \approx E_i d^{-\frac{1}{m}} \quad \text{where} \quad m = \omega_i + 1. \tag{33}$$

Equation (33) means that the relation between E_{BD} and d conforms to a minus power relation with a power exponent of $1/m$ as the dielectric thickness increases. In addition, the smallest m is 4 in Equation (33) or the largest a is $1/4$ for Equation (12) based on Figure 12.

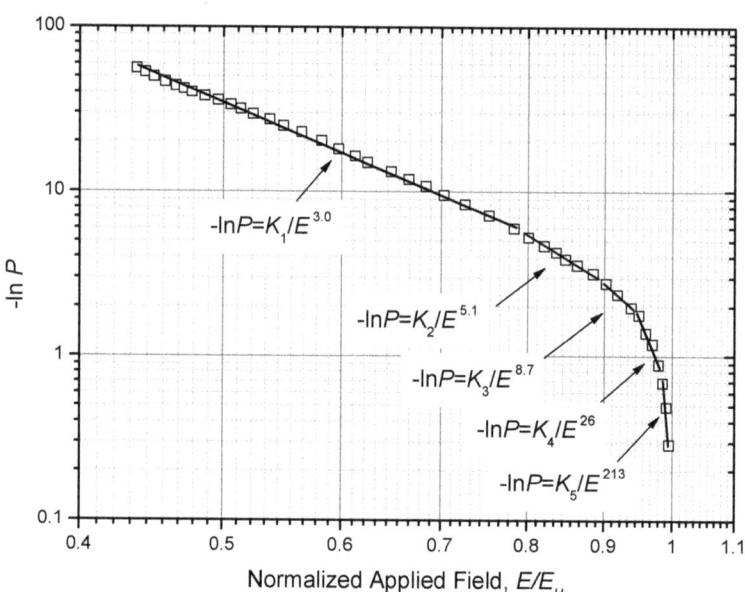

Figure 11. Plot of the effect of $-\ln P$ on E/E_H in a log-log coordinate system.

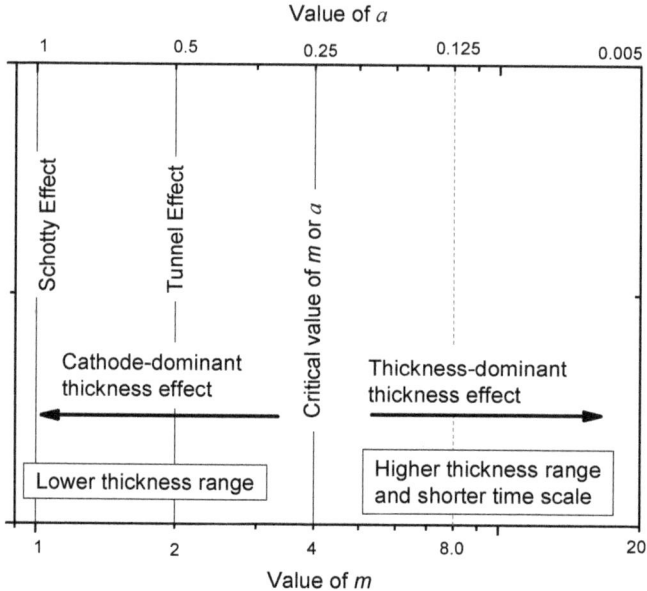

Figure 12. Different breakdown mechanisms for different values of m betrayed by the minus power relation.

By generalizing the three cases of solutions in Equation (25), Equation (28) and Equation (33) together, it can be concluded that the thickness effect of solid dielectric meets the minus power relation—that is:

$$E_{BD} = E_k d^{-1/m}. \tag{34}$$

where E_k and m are both constants. When $m = 1$ or $a = 1$, this relation represents a breakdown mechanism related to the Schottky effect; when $m = 2$ or $a = 1/2$, this relation represents a breakdown mechanism related to the tunnel effect; when $m \geq 4$ or $a \leq 1/4$, the breakdown mechanism is related to the avalanche process. Figure 12 shows the three types of breakdown mechanism for the minus power relation with m as the argument.

5. Minus Power Relation from Weibull Statistics

The above analysis is just from the perspective of the breakdown mechanism. It is also necessary to analyze the thickness effect from the perspective of statistical distribution.

5.1. Deduction for the Minus Power Relation

The Weibull distribution is a widely-used method for analyzing failure events [63], especially for the breakdown in insulation dielectrics [64–68]. The two-parameter Weibull distribution is as follows:

$$F(E) = 1 - \exp\left(-\frac{E^m}{\eta}\right), \tag{35}$$

where $F(E)$ is the breakdown probability, E is the applied field; m and η are the shape parameter and the dimension parameter, respectively. If E is equal to $\eta^{1/m}$, $F(E) = 0.6321$. Moreover, this field is defined as E_{BD}. Now, assume that the reliability of a solid dielectric with a thickness of d_1 is R and this sample is placed in a field of E. Since $F + R = 1$, R can be expressed as follows:

$$R(E) = \exp\left(-\frac{E^m}{\eta}\right). \tag{36}$$

Further, assume that M samples with the same thickness and configuration are placed in series in the field of E [69].

By neglecting the electrode effect and edge effect and by assuming that the reliability of each sample is equal, the following relation would hold true:

$$R_M(E) = \exp\left(-\frac{E^m}{\eta}\right)^M = \exp\left(-\frac{E^m}{\eta/M}\right). \tag{37}$$

Then, the breakdown probability of the thick sample is:

$$F_M(E) = 1 - R_M(E) = 1 - \exp\left(-\frac{E^m}{\eta/M}\right). \tag{38}$$

Based on the definition of E_{BD} from the perspective of the Weibull distribution, the breakdown strength E_{BDM} for the thick sample is:

$$E_{BDM} = \left(\frac{\eta}{M}\right)^{\frac{1}{m}} = \left(\frac{1}{M}\right)^{\frac{1}{m}} \cdot \eta^{\frac{1}{m}}. \tag{39}$$

Taking into account the assumptions that $d_M = Md_1$ and $\eta^{1/m} = E_{BD1}$, Equation (39) is simplified as:

$$E_{BDM} = \frac{d_1^{\frac{1}{m}} E_{BD1}}{d_M^{\frac{1}{m}}}. \tag{40}$$

If the thin samples are standard with a unit thickness (for example, $d_1 = 1$ mm), then $d_1^{1/m} E_{BD1} = E_1$. Getting rid of the subscript of M in Equation (40) gives:

$$E_{BD} = E_1 d^{-\frac{1}{m}}. \tag{41}$$

Since $1/m > 0$, E_{BD} will decrease as d increases. Equation (41) is exactly the same as that derived in Equation (34). This is what is betrayed from the perspective of the Weibull distribution.

Making a logarithmic transformation for (41) and letting $C = \lg E_{BD0}$ give:

$$\lg E_{BD} = C_1 - \frac{1}{m} \lg d. \tag{42}$$

Equation (42) means that $\lg E_{BD}$ is linear to $\lg d$ with a slope rate of $-1/m$. Thus, the value of m can be known conveniently by fitting the E_{BD} vs. d data linearly in a log-log coordinate system.

Figure 13 shows the fitting results of two types of PMMA samples under nanosecond pulses [70]; one is pure, the other is porous. It can be seen that the m of the pure PMMA is large (7.4), but the m of the porous PMMA is small (3.8). It is worth mentioning that an m of 3.8 is close to the theoretical value of m = 4 in the last section. It is also worth mentioning that the value of m is affected by the dielectric quality obviously. Based on the research in [70], the better the dielectric quality is, the larger m is.

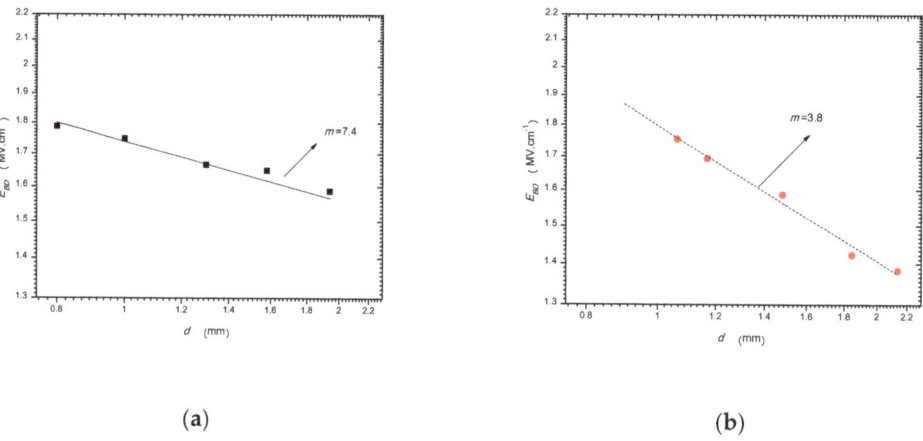

Figure 13. Fitting of different groups of E_{BD} v.s. d data from two types of PMMA samples. (**a**) Pure PMMA; (**b**) porous PMMA.

5.2. Expectation and Standard Error of Weibull Distribution

The value of m not only reflects the dielectric quality, but also the breakdown characteristics. In Equations (41) and (42), m is defined as the shape parameter of the Weibull distribution. Different values of m mean different Weibull distributions. The breakdown probability density of the Weibull distribution in Equation (35) is:

$$f(E) = \frac{mE^{m-1}}{\eta} \exp\left(-\frac{E^m}{\eta}\right). \tag{43}$$

Figure 14a shows the $f(E)$ for different values of m. From this figure, it can be seen that the larger the value of m is, the more concentrated the distribution is.

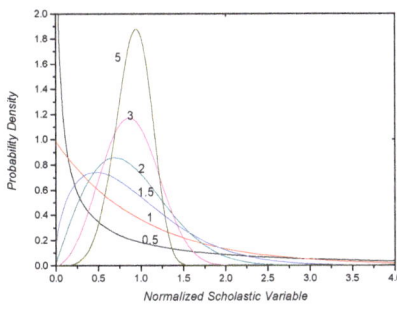
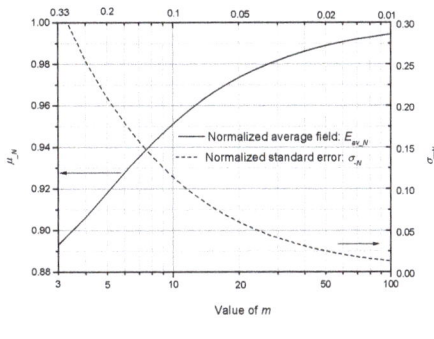

(a)   (b)

Figure 14. Weibull distribution for different values of m. (a) Breakdown probability density v.s. scholastic variable for different m; (b) normalized expectation and standard error of Weibull distribution dependent on m.

The expectation μ and standard error σ of the Weibull distribution in Equation (35) are:

$$\mu(m) = \eta^{\frac{1}{m}} \Gamma\left(1 + \frac{1}{m}\right), \tag{44}$$

and

$$\sigma(m) = \eta^{\frac{1}{m}} \left[\Gamma\left(1 + \frac{2}{m}\right) - \Gamma^2\left(1 + \frac{1}{m}\right)\right]^{\frac{1}{2}}, \tag{45}$$

where Γ means the gamma function. The normalized μ_N and σ_N, which are divided by $\eta^{1/m}$, are shown in Figure 14b. From this figure, it is seen that the larger m is, the larger μ_N is(or closer μ_N to 1) and the smaller σ_N is (or closer σ_N to 0).

5.3. Physical Meaning of a or 1/m

By comparing the deduced minus power relation in Equation (41) from the perspective of the breakdown mechanism and that in Equation (12) obtained by fitting the experimental result, one can easily obtain:

$$a = 1/m. \tag{46}$$

Now, what is the practical meaning of a or $1/m$ in the minus power relation for the thickness effect of solid dielectrics? In order to answer this question, the unified standard error σ' is defined, which is as follows:

$$\sigma' = \sigma/\mu, \tag{47}$$

This value is similar to the standard error in the normal distribution. Figure 15a shows the comparison of σ' with a; from the figure, it is seen that the two values are basically equal to each other, only with a deviation, δ, smaller than 0.03, as shown in the inset in Figure 15b. Here, δ is defined as $(\sigma' - a)$. This means that the minus power exponent a can be represented by the standard error of the E_{BD} on a fixed thickness. In other word, a has the physical meaning of the standard error of E_{BD} for a fixed thickness.

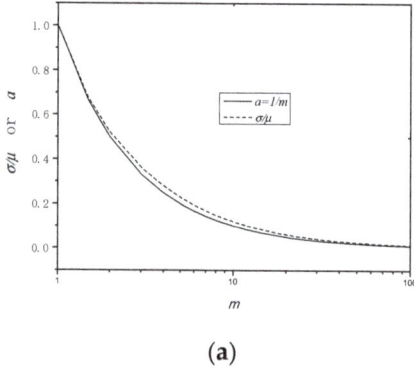 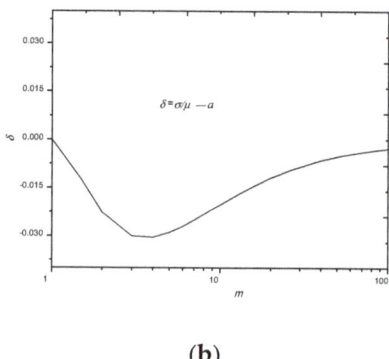

(a) (b)

Figure 15. Comparison of the standard error σ/μ with the value of a (or $1/m$). (**a**) σ/μ and a (or $1/m$) are dependent on m; (**b**) δ (=$\sigma/\mu - a$) is dependent on m.

As a conclusion for this section, the minus power relation can also be deduced from the Weibull distribution. The shape parameter m reflects the dielectric quality. The larger m is, the better the quality a dielectric has. In addition, the power exponent $1/m$ or a has the physical meaning of the standard error of E_{BD}.

6. Discussion on the Power Exponent of a or $1/m$ in Minus Power Relation

Here, two groups of discussion are made: one is for factors influencing a or $1/m$, the other is for the information betrayed from a or $1/m$ once it is obtained.

6.1. Factors Influencing a

In Table 3, different values of a in the minus power relation from experiments in different conditions are summarized. Based on this table, factors such as thickness range, time scale, temperature, electrode configuration, and environmental liquid can be discussed.

6.1.1. Temperature

Helgee researched the thickness effect under different temperatures [47]. By fitting each group of E_{BD} v.s. d data, the value of a under different temperatures T can be obtained and plotted, as shown in Figure 16.

This figure reveals that a increases slightly as T increases in a range of $a \geq 0.25$ (or $m \leq 4$). It is believed that this kind of dependency can be explained from the perspective of the breakdown model of Forlani.

Based on the analysis in Section 4.2, the cathode is involved in the breakdown process when $m \leq 4$. In addition, the electron injection mechanism is probably related to the tunnel effect when $2 < m \leq 4$. According to the field-induced emission formula affected by temperature:

$$j_i(T) \approx AE^2 \exp\left(-\frac{B\theta(y)}{E}\right)\left[1 + \left(\frac{\pi k_B T}{D}\right)^2\right], \quad (48)$$

where D is the transmittance coefficient. It is seen that the injection ability of cathode is enhanced as T increases, which means that the role the cathode plays in the breakdown process becomes more obvious. Based on Figure 12, if the role of cathode becomes obvious, the value of a should increase.

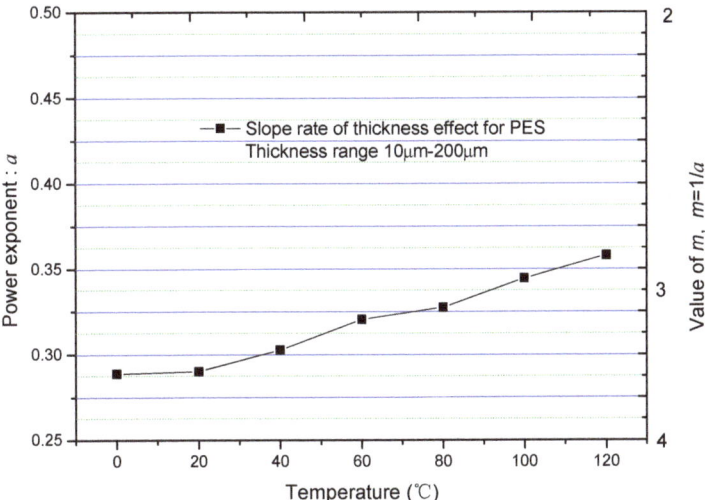

Figure 16. The dependence of a on temperature.

6.1.2. Time Scale

When the time scale of the applied voltage increases, the heat accumulated in the dielectric become more obvious and the breakdown mechanism gradually changes from an electronic process to a thermal process [71]. If the heat is accumulated more easily, the temperature of the cathode would increase. According to the dependence of a on T in Figure 16, a would increase. By reviewing the relevant literatures, some proofs are found. For example, a is 0.125 under 10 ns by L. Zhao et al. [61] and a is 0.7 under dc voltage by J. H. Mason [44]. As to these two groups of experiments, the test samples are all made of PE, the sample thickness falls in a millimeter range and the samples are all immersed in oil. Thus, they can be compared together.

6.1.3. Electrode Configuration and Ambient Liquid

Based on Table 3, the effects of the electrode configuration and the ambient liquid on a can also be discussed. From the lines of 1991/Mason on PP with a thickness of 8–76 μm, it is seen that a more uniform electric field would result in a smaller value of a. From the lines of 1991/Mason on PVC with a thickness of 40–500 μm, it is seen that a higher ε_r of the ambient liquid may result in a lower value of a. These two phenomena can be explained from the perspective of the practical meaning of a. When the field is more uniform, the distribution of E_{BD} on a fixed thickness would be more concentrated, since the influence causing the divergence of the standard error becomes weak. Similarly, when ε_r of the ambient liquid is high, the field tends to focus on a dielectric which has a lower ε_r—i.e., the test sample. Thus, the influence of the ambient liquid causing the divergence of the standard error of E_{BD} becomes less, and a becomes smaller.

6.1.4. Thickness Range

In Figure 5, the distribution of a in a wide thickness range is plotted. This figure shows that the dependency of a on the dielectric thickness d seems "random". It is believed that two main mechanisms may be responsible for this "random distribution". Firstly, when the thickness gets smaller, the breakdown process is affected more seriously by the cathode, such as the Al_2O_3 film layers embedded in metal-oxide-semiconductor ~MOS [72,73]. Thus, the value of a tends to increase based on Figure 12. Secondly, when the thickness gets larger, heat can easily accumulate in the dielectric and the temperature of the cathode would increase. Thus, the value of a also tends to increase based on Figure 16. Aside from these two effects, other factors such as the time scale, the temperature, the electrode configuration, and the ambient liquid can all affect the specific value of a in a fixed group of experiments. Thus, a "random distribution" on a takes on.

As a conclusion of this subsection, factors such as time scale, temperature, electric uniformity, ambient liquid, and thickness range can all have influence on the value of a. Test conditions such as short time scale, low temperature, uniform field, and high ε_r of ambient liquid can all result in a small value of a.

6.2. What Does m Betray When m > 4 (or a < 0.25)?

From Table 3, it is noted that the results of $a < 0.25$ are relatively less. $a < 0.25$ is equivalent to $m > 4$. For convenience, the case of $m > 4$ is discussed.

From Figure 12, it is seen that when $m < 4$ and the dielectric thickness is lower, the mechanism of the thickness effect is mostly related to the electron injection of the cathode. When $m \geq 4$ and the dielectric thickness is larger, the mechanism of the thickness effect is mostly related to the electron avalanche formation. As to solid insulation structures, the thickness is usually from mm to cm range. Thus, discussion of the case of $m \geq 4$ has more practical value. Here, this case is discussed especially.

Now, a question comes: once the value of m for a thick insulation material or structure is obtained, what kind of information can be betrayed? This question can be answered from three perspectives.

(1) From the perspective of mechanism: From Figure 11, it is seen that the larger the value of m is, the higher the E_{BD} is. In addition, from Figure 10 it is seen that a higher E_{BD} corresponds to a higher avalanche probability P. This means that the formation of the avalanche is easier for a dielectric with a larger m than that with a smaller m.

(2) From the perspective of dielectric quality: from Figure 13 it is seen that a larger m corresponds to a better dielectric quality.

(3) From the perspective of breakdown distribution: From Figure 14a, it is seen that the larger the value of m is, the more concentrated the breakdown probability density is. In addition, from Figure 14b, a larger m corresponds to a higher expectation and a smaller standard error, which means that the E_{BD} is also higher.

Now, let us consider that this question from the perspective of a. When $a < 0.25$, the smaller the value of a is, the more easily the avalanche forms, and the higher the E_{BD} is, the better the dielectric quality and the more concentrated the breakdown events are. Table 4 summarizes all the information betrayed when $m \geq 4$ or $a \leq 0.25$.

Table 4. Information betrayed from m when $m \geq 4$ for thick dielectrics (or information betrayed from a when $a \leq 0.25$).

Perspective	A Larger m (or a Smaller a) Means:
Breakdown mechanism	A larger P and a larger E_{BD}.
Dielectric purity	A better dielectric purity.
Statistics	A higher E_{BD} and a smaller σ

7. Conclusions

Based on the review and the analysis in the whole paper, the questions put forward in the introduction can tentatively be answered:

(1) The minus power relation is preferable to characterize the thickness effect of solid dielectrics.

(2) The physical mechanism responsible for the minus power relation of the thickness effect lies in the electron injection of the cathode combined with the avalanche process in dielectrics.

(3) The application range of the minus power relation is from several Å s to several millimeters.

In addition, the following conclusions have been drawn:

(4) The practical meaning of the minus power exponent a is the relative standard error of the distribution (σ/μ) of E_{BD} on a fixed thickness.

(5) The value of a (or $1/m$) is different in different experiments, which is affected by factors such as time scale, temperature, electric field uniformity, ambient liquid, and thickness range. The specific value of a or m needs to be tested from experiments.

(6) A smaller value of a corresponds to an easy avalanche formation, a higher E_{BD}, a better dielectric quality, and a more concentrated E_{BD} distribution.

Author Contributions: C.L.L. supplied a lot of precious references for L.Z. to conduct the review on the thickness effect of solid dielectric. L.Z. wrote the paper. All authors have read and agreed to the published version of the manuscript.

Funding: This work was supported by the National Natural Science Foundation of China under Grant 51377135.

Conflicts of Interest: The authors declare no conflict of interest.

References

1. Korovin, S.D.; Rostov, V.V.; Polevin, S.D.; Pegel, I.V.; Schamiloglu, E.; Fuks, M.I.; Barker, R.J. Pulsed power-driven high-power microwave sources. *Proc. IEEE* **2004**, *92*, 1082–1095. [CrossRef]
2. Roth, I.S.; Sincerny, P.S.; Mandelcorn, L.; Mendelsohn, M.; Smith, D.; Engel, T.G.; Schlitt, L.; Cooke, C.M. Vacuum insulator coating development. In Proceedings of the 11th IEEE International Pulsed Power Conference, Baltimore, MD, USA, 29 June 1997; IEEE: Piscataway, NJ, USA, 1997; pp. 537–542.
3. Stygar, W.A.; Lott, J.A.; Wagoner, T.C.; Anaya, V.; Harjes, H.C. Improved design of a high-voltage vacuum-insulator interface. *Phys. Rev. Spec. Top. Accel. Beams* **2005**, *8*, 050401. [CrossRef]
4. Zhao, L.; Peng, J.C.; Pan, Y.F.; Zhang, X.B.; Su, J.C. Insulation Analysis of a Coaxial High-Voltage Vacuum Insulator. *IEEE Trans. Plasma Sci.* **2010**, *38*, 1369–1374. [CrossRef]
5. Mesyats, G.A.; Korovin, S.D.; Gunin, A.V.; Gubanov, V.P.; Stepchenko, A.S.; Grishin, D.M.; Landl, V.F.; Alekseenko, P.I. Repetitively pulsed high-current accelerators with transformer charging of forming lines. *Laser Part. Beams* **2003**, *21*, 198–209. [CrossRef]
6. Oakes, W.G. The intrinsic electric strength of polythene and its variation with temperature. *J. Inst. Electr. Eng.* **1948**, *95*, 36–44. [CrossRef]
7. von Hippel, A. Electric Breakdown of Solid and Liquid Insulators. *J. Appl. Phys.* **1937**, *8*, 815–832. [CrossRef]
8. von Hippel, A. Electronic conduction in insulating crystals under very high field strength. *Phys. Rev.* **1938**, *54*, 1096–1102. [CrossRef]
9. von Hippel, A.; Davisson, J.W. The propagation of electron waves in ionic single crystals. *Phys. Rev.* **1940**, *57*, 156–157. [CrossRef]
10. von Hippel, A.; Lee, G.M. Scattering, trapping, and release of electrons in NaCl and in mixed crystals of NaCl and AgCl. *Phys. Rev.* **1941**, *59*, 824–826. [CrossRef]
11. von Hippel, A.; Maurer, R.J. Electric breakdown of glasses and crystals as a function of temperature. *Phys. Rev.* **1941**, *59*, 820–823. [CrossRef]
12. von Hippel, A.; Schulman, J.H.; Rittner, E.S. A New Electrolytic Selenium Photo-Cell. *J. Appl. Phys.* **1946**, *17*, 215–224. [CrossRef]
13. von Hippel, A.V.; Auger, R.S. Breakdown of Ionic Crystals by Electron Avalanches. *Phys. Rev.* **1949**, *76*, 127–133. [CrossRef]
14. von Hippel, A.R. From ferroelectrics to living systems. *IEEE Trans. Electron. Dev.* **1969**, *16*, 602. [CrossRef]
15. Frohlich, H. Theory of electrical breakdown in ionic crystals. In *Proceedings of Royal Society of London*; Series A: Mathematical and Physical Sciences; Royal Society: London, UK, 1937; Volume 160, pp. 230–241.
16. Frohlich, H. Theory of electrical breakdown in ionic crystals. II. In *Proceedings of Royal Society of London*; Series A: Mathematical and Physical Sciences; Royal Society: London, UK, 1939; Volume 172, pp. 94–106.
17. Frohlich, H. Dielectric breakdown in ionic crystals. *Phys. Rev.* **1939**, *56*, 349–352. [CrossRef]
18. Frohlich, H. Electric breakdown of ionic crystals. *Phys. Rev.* **1942**, *61*, 200–201. [CrossRef]
19. Frohlich, H. Theory of dielectric breakdown. *Nature* **1943**, *151*, 339–340. [CrossRef]
20. Frohlich, H. On the Theory of Dielectric Breakdown in Solids. In *Proceedings of Royal Society of London*; Series A: Mathematical and Physical Sciences; Royal Society: London, UK, 1947; pp. 521–532.
21. Frohlich, H.; Seitz, F. Notes on the Theory of Dielectric Breakdown in Ionic Crystals. *Phys. Rev.* **1950**, *79*, 526–527. [CrossRef]

22. Seitz, F.; Johnson, R.P. Modern Theory of Solids. I. *J. Appl. Phys.* **1937**, *8*, 84–97. [CrossRef]
23. Seitz, F.; Johnson, R.P. Modern Theory of Solids. III. *J. Appl. Phys.* **1937**, *8*, 246–260. [CrossRef]
24. Seitz, F.; Johnson, R.P. Modern Theory of Solids. II. *J. Appl. Phys.* **1937**, *8*, 186–199. [CrossRef]
25. Seitz, F.; Read, T.A. Theory of the Plastic Properties of Solids. I. *J. Appl. Phys.* **1941**, *12*, 100–118. [CrossRef]
26. Seitz, F.; Read, T.A. Theory of the Plastic Properties of Solids. III. *J. Appl. Phys.* **1941**, *12*, 470–486. [CrossRef]
27. Seitz, F.; Read, T.A. Theory of the Plastic Properties of Solids. II. *J. Appl. Phys.* **1941**, *12*, 170–186. [CrossRef]
28. Seitz, F.; Read, T.A. Theory of the Plastic Properties of Solids. IV. *J. Appl. Phys.* **1941**, *12*, 538–554. [CrossRef]
29. Stark, K.H.; Garton, G.C. Electrical Strength of Irradiation Polymers. *Nature* **1955**, *176*, 1225–1226. [CrossRef]
30. Austen, A.E.W.; Whitehead, S. The electric strength of some solid dielectrics. In *Proceedings Royal Society in London*; Royal Society: London, UK, 1940; Volume 176, pp. 33–50.
31. Whitehead, S. *Dielectric Breakdown of Solid*; Clarendon Press: Oxford, UK, 1951.
32. O'Dwyer, J.J. Dielectric breakdown in solid. *Adv. Phys.* **1958**, *7*, 349–394. [CrossRef]
33. O'Dwyer, J.J. The theory of avalanche breakdown in solid dielectrics. *J. Phys. Chem. Solids* **1967**, *7*, 1137–1144. [CrossRef]
34. O'Dwyer, J.J. Theory of High Field Conduction in a Dielectric. *J. Appl. Phys.* **1969**, *40*, 3887–3890. [CrossRef]
35. Zurlo, G.; Destrade, M.; DeTommasi, D.; Puglisi, G. Catastrophic Thinning of Dielectric Elastomers. *Phys. Rev. Lett.* **2017**, *118*, 078001. [CrossRef]
36. Christensen, L.R.; Hassager, O.; Skov, A.L. Electro-Thermal and -Mechanical Model of Thermal Breakdown in Multilayered Dielectric Elastomers. *AIChE J.* **2020**, *66*. [CrossRef]
37. Vermeer, J. The impulse breadown strength of pyrex glass. *Phys. X X* **1954**, *20*, 313–326.
38. Seitz, F. On the Theory of Electron Multiplication in Crystals. *Phys. Rev.* **1949**, *76*, 1376–1393. [CrossRef]
39. Cooper, R.; Rowson, C.H.; Waston, D.B. Intrinsic electric strength of polythene. *Nature* **1963**, *197*, 663–664. [CrossRef]
40. Forlani, F.; Minnaja, N. Thickness influence in breakdown phenomena of thin dielectric films. *Phys. Stat. Sol.* **1964**, *4*, 311–324. [CrossRef]
41. Cooper, R. The dielectric strength of solid dielectrics. *Br. J. Appl. Phys.* **1966**, *17*, 149–166. [CrossRef]
42. Merrill, R.C.; West, R.A. A New Dried Anodized Aluminum Capacitor. *IEEE Trans. Parts Mater. Packag.* **1965**, *1*, 224–229. [CrossRef]
43. Mason, J.H. Breakdown of insulation by discharges. *Proc. IEE Part IIA Insul. Mater.* **1953**, *100*, 149–158. [CrossRef]
44. Mason, J.H. Effects of thickness and area on the electric strength of polymers. *IEEE Trans. Electr. Insul.* **1991**, *26*, 318–322. [CrossRef]
45. Mason, J.H. Effects of frequency on the electric strength of polymers. *IEEE Trans. Electr. Insul.* **1992**, *27*, 1213–1216. [CrossRef]
46. Yilmaz, G.; Kalenderli, O. The effect of thickness and area on the electric strength of thin dielectric films. In Proceedings of the IEEE International Symposium on Electrical Insulation, Montreal, QC, Canada, 16–19 June 1996; pp. 478–481.
47. Yilmaz, G.; Kalenderli, O. Dielectric behavior and electric strength of polymer films in varying thermal conditions for 5 Hz to 1 MHz frequency range. In Proceedings of the Electrical Insulation Conference and Electrical Manufacturing and Coil Winding, Rosemont, IL, USA, 22–25 September 1997; pp. 269–271.
48. Singh, A.; Pratap, R. AC electrial breakdown in thin magnesium oxide. *Thin Solid Films* **1982**, *87*, 147–150. [CrossRef]
49. Singh, A. Dielectric breakdown study of thin La_2O_3 Films. *Thin Solid Films* **1983**, *105*, 163–168. [CrossRef]
50. Ye, Y.; Zhang, S.C.; Dogan, F.; Schamiloglu, E.; Gaudet, J.; Castro, P.; Roybal, M.; Joler, M.; Christodoulou, C. Influence of nanocrystalline grain size on the breakdown strength ceramic dielectrics. In *Digest of Technical Papers, PPC-2003. 14th IEEE International Pulsed Power Conference*; IEEE: Piscataway, NJ, USA, 2003; pp. 719–722.
51. Yoshino, K.; Harada, S.; Kyokane, J.; Inuishi, Y. Electrical properties of hexatriacontane single crystal. *J. Phys. D Appl. Phys.* **1979**, *12*, 1535–1540. [CrossRef]
52. Theodosiou, K.; Vitellas, I.; Gialas, I.; Agoris, D.P. Polymer film degradation and breakdown in high voltage ac fields. *J. Electr. Eng.* **2004**, *55*, 225–231.
53. Chen, G.; Zhao, J.; Li, S.; Zhong, L. Origin of thickness dependent dc electrical breakdown in dielectrics. *Appl. Phys. Lett.* **2012**, *100*, 222904.

54. Agwal, V.K.; Srivastava, V.K. Thickness dependence of breakdown filed in thin films. *Thin Solid Films* **1971**, *1971*, 377–381. [CrossRef]
55. Forlani, F.; Minnaja, N. Electrical Breakdown in Thin Dielectric Films. *J. Vac. Sci. Technol.* **1969**, *6*, 518–525. [CrossRef]
56. Mason, J.H. Breakdown of Solid Dielectrics in Divergent Fields. *Proc. IEE Part C Monogr.* **1955**, *102*, 254–263. [CrossRef]
57. Mason, J.H. Comments on 'Electric breakdown strength of aromatic polymers-dependence on film thickness and chemical structure'. *IEEE Trans. Electr. Insul.* **1992**, *27*, 1061. [CrossRef]
58. Helgee, B.; Bjellheim, P. Electric breakdown strength of aromatic polymers: Dependence on film thickness and chemical structure. *IEEE Trans. Electr. Insul.* **1991**, *26*, 1147–1152. [CrossRef]
59. Yilmaz, G.; Kalenderli, O. Dielectric Properties of Aged Polyester Films. In Proceedings of the IEEE Annual Report—Conference on Electrical Insulation and Dielectric Phenomena, Minneapolis, MN, USA, 19–22 October 1997; pp. 444–446.
60. Diaham, S.; Zelmat, S.; Locatelli, M.L.; Dinculescu, S.; Decup, M.; Lebey, T. Dielectric Breakdown of Polyimide Films: Area, Thickness and Temperature Dependence. *IEEE Trans. Dielectr. Electr. Insul.* **2010**, *17*, 18–27. [CrossRef]
61. Zhao, L.; Liu, G.Z.; Su, J.C.; Pan, Y.F.; Zhang, X.B. Investigation of Thickness Effect on Electric Breakdown Strength of Polymers under Nanosecond Pulses. *IEEE Trans. Plasma Sci.* **2011**, *39*, 1613–1618. [CrossRef]
62. Neusel, B.C.; Schneider, G.A. Size-dependence of the dielectric breakdown strength from nano- to millimeter scale. *J. Mech. Phys. Solids* **2013**, *63*, 201–213. [CrossRef]
63. Weibull, W. A statistical distribution function of wide applicability. *J. Appl. Mech.* **1951**, *18*, 293–297.
64. Kiyan, T.; Ihara, T.; Kameda, S.; Furusato, T.; Hara, M.; Akiyama, H. Weibull Statistical Analysis of Pulsed Breakdown Voltages in High-Pressure Carbon Dioxide Including Supercritical Phase. *IEEE Trans. Plasma Sci.* **2011**, *39*, 1792. [CrossRef]
65. Dissado, L.A.; Fothergill, J.C.; Wolfe, S.V.; Hill, R.M. Weibull Statistics in Dielectric Breakdown: Theoretical Basis, Applications and Implications. *IEEE Trans. Electr. Insul.* **1984**, *EI-19*, 227–233. [CrossRef]
66. Tsuboi, T.; Takami, J.; Okabe, S.; Inami, K.; Aono, K. Weibull Parameter of Oil-immersed Transformer to Evaluate Insulation Reliability on Temporary Overvoltage. *IEEE Trans. Dielectr. Electr. Insul.* **2010**, *17*, 1863–1868. [CrossRef]
67. Wilson, M.P.; Given, M.J.; Timoshkin, I.V.; MacGregor, S.J.; Wang, T.; Sinclair, M.A.; Thomas, K.J.; Lehr, J.M. Impulse-driven Surface Breakdown Data: A Weibull Statistical Analysis. *IEEE Trans. Plasma Sci.* **2012**, *40*, 2449–2457. [CrossRef]
68. Hauschild, W.; Mosch, W. *Statistical Techniques for High Voltage Engineering*; IET: London, UK, 1992.
69. Zhao, L.; Su, J.C.; Zhang, X.B.; Pan, Y.F.; LI, R.; Zeng, B.; Cheng, J.; Yu, B.X.; Wu, X.L. A Method to design composite insulation structures based on reliability for pulsed power systems. *Laser Part. Beams* **2014**, *32*, 1–8. [CrossRef]
70. Zhao, L.; Su, J.C.; Pan, Y.F.; Li, R.; Zeng, B.; Cheng, J.; Yu, B.X. Correlation between volume effect and lifetime effect of solid dielectrics on nanosecond time scale. *IEEE Trans. Dielectr. Electr. Insul.* **2014**, *22*, 1769–1776. [CrossRef]
71. Zhao, L. Theoretical Calculation on Formative Time Lag in Polymer Breakdown on a Nanosecond Time Scale. *IEEE Trans. Dielectr. Electr. Insul.* **2020**, *27*, 1051–1058. [CrossRef]
72. Ludeke, R.; Cuberes M., T.; Cartier, E. Hot carrier transport effects in Al_2O_3-based metal-oxide-semiconductor structures. *J. Vac. Sci. Technol. B Microelectron. Nanom. Struct.* **2000**, *18*, 2153–2159. [CrossRef]
73. Ludeke, R.; Cuberes, M.T.; Cartier, E. Local transport and trapping issues in Al2O3 gate oxide structures. *Appl. Phys. Lett.* **2000**, *76*, 2886–2888. [CrossRef]

Publisher's Note: MDPI stays neutral with regard to jurisdictional claims in published maps and institutional affiliations.

© 2020 by the authors. Licensee MDPI, Basel, Switzerland. This article is an open access article distributed under the terms and conditions of the Creative Commons Attribution (CC BY) license (http://creativecommons.org/licenses/by/4.0/).

Article

Porous Polydimethylsiloxane Elastomer Hybrid with Zinc Oxide Nanowire for Wearable, Wide-Range, and Low Detection Limit Capacitive Pressure Sensor

Gen-Wen Hsieh [1,*], Liang-Cheng Shih [2] and Pei-Yuan Chen [2]

1. Institute of Lighting and Energy Photonics, College of Photonics, National Yang Ming Chiao Tung University, 301, Section 2, Gaofa 3rd Road, Guiren District, Tainan 71150, Taiwan
2. Institute of Photonic System, College of Photonics, National Yang Ming Chiao Tung University, 301, Gaofa 3rd Road, Section 2, Guiren District, Tainan 71150, Taiwan; lighttime0625@yahoo.com.tw (L.-C.S.); aazz55255361@gmail.com (P.-Y.C.)
* Correspondence: cwh31@nctu.edu.tw or cwh31@nycu.edu.tw; Tel.: +86-(0)-6303-2121 (ext. 57797); Fax: +86-(0)-6303-2535

Abstract: We propose a flexible capacitive pressure sensor that utilizes porous polydimethylsiloxane elastomer with zinc oxide nanowire as nanocomposite dielectric layer via a simple porogen-assisted process. With the incorporation of nanowires into the porous elastomer, our capacitive pressure sensor is not only highly responsive to subtle stimuli but vigorously so to gentle touch and verbal stimulation from 0 to 50 kPa. The fabricated zinc oxide nanowire–porous polydimethylsiloxane sensor exhibits superior sensitivity of 0.717 kPa^{-1}, 0.360 kPa^{-1}, and 0.200 kPa^{-1} at the pressure regimes of 0–50 Pa, 50–1000 Pa, and 1000–3000 Pa, respectively, presenting an approximate enhancement by 21−100 times when compared to that of a flat polydimethylsiloxane device. The nanocomposite dielectric layer also reveals an ultralow detection limit of 1.0 Pa, good stability, and durability after 4000 loading–unloading cycles, making it capable of perception of various human motions, such as finger bending, calligraphy writing, throat vibration, and airflow blowing. A proof-of-concept trial in hydrostatic water pressure sensing has been demonstrated with the proposed sensors, which can detect tiny changes in water pressure and may be helpful for underwater sensing research. This work brings out the efficacy of constructing wearable capacitive pressure sensors based on a porous dielectric hybrid with stress-sensitive nanostructures, providing wide prospective applications in wearable electronics, health monitoring, and smart artificial robotics/prosthetics.

Keywords: capacitive pressure sensor; porous polydimethylsiloxane; stress-sensitive; wearable electronic; zinc oxide nanowire

Citation: Hsieh, G.-W.; Shih, L.-C.; Chen, P.-Y. Porous Polydimethylsiloxane Elastomer Hybrid with Zinc Oxide Nanowire for Wearable, Wide-Range, and Low Detection Limit Capacitive Pressure Sensor. *Nanomaterials* 2022, 12, 256. https://doi.org/10.3390/nano12020256

Academic Editor: Teresa Cuberes

Received: 13 December 2021
Accepted: 11 January 2022
Published: 14 January 2022

Publisher's Note: MDPI stays neutral with regard to jurisdictional claims in published maps and institutional affiliations.

Copyright: © 2022 by the authors. Licensee MDPI, Basel, Switzerland. This article is an open access article distributed under the terms and conditions of the Creative Commons Attribution (CC BY) license (https://creativecommons.org/licenses/by/4.0/).

1. Introduction

To meet the growing demand for wearable healthcare electronics and human–machine interfaces, nanocomposite materials that employ flexible polymers in conjunction with stimuli-sensitive nanostructures have engrossed significant attention for the consciousness of temperature [1–5], moisture [6,7], light [8,9], and touch [10–13]. Unlike conventional composites or a sole type of material, polymer-based nanocomposites may offer considerable enhancement in thermal, electrical, optical, chemical, or mechanical properties. Among them, a variety of nanostructures with highly conductive [14–16], dielectric [17–19], piezoelectric [20–22], triboelectric [23–25], photo-responsive [26–28], or stress-sensitive [10,29–31] properties have been studied for flexible pressure sensors because of their potential in amplifying stress/strain sensing capability for human motion detection, health diagnosis, and electronic skin.

Recently, capacitive pressure sensing has been intensively investigated, owing to its simple geometry, low power consumption, and good environmental stability [18].

In a simple parallel plate capacitor, capacitance is commonly changed as a function of external pressure. In particular, elastomeric polydimethylsiloxane (PDMS)-based silicone rubbers have been chosen as the dielectric layer for capacitive pressure sensors due to their superior flexibility, nontoxicity, and low material cost [32,33]. However, restricted by their viscoelastic property and low compressibility, such PDMS films are not able to produce enough deformation upon very small pressures. Moreover, after pressure unloading, their recovery time of return to initial condition is rather slow [18]. To overcome these limitations, adoption of micro-nano-scaled structures or pores into an elastomeric matrix is a possible means to improve sensing performance [34–38]. With applied pressure, these embedded air gaps or pores in the deformed PDMS films can induce massive volumetric deformation as well as increments in effective dielectric permittivity, which can in turn increase the capacitance change and pressure sensitivity. However, such enhancements can only be achieved in the low-pressure regime; when these air gaps or pores are nearly squeezed under heavy load, the flattened PDMS becomes hardly compressible. Additionally, the fabrication processes for these flexible pressure sensors are complicated, high-cost, and difficult to control, which is not practical for human tactile interactions.

In this regard, an appropriate way to generate high-performance capacitive pressure sensors is to tailor low-dimensional nanostructures with porous or microstructured elastomeric polymers by merging their functionalities. Owing to the high surface area to volume ratio, several types of nanostructures can promote the formation of a large interfacial area between polymer and nanofiller, which is beneficial to a higher level of polarization. For instance, Mu et al. demonstrated a nanocomposite dielectric layer of $CaCu_3Ti_4O_{12}$ ceramic nanocrystals (with high dielectric permittivity) and porous PDMS matrix (with low dielectric permittivity) [39], presenting low compressibility and high sensitivity compared to that of pure PDMS. Pruvost et al. produced a composite dielectric foam decorated with conductive carbon black particles on the inner surface of pores [40], which can enhance the effective dielectric permittivity and the capacitance change under applied stress. Moreover, Kou et al. proposed a wireless flexible pressure sensor containing a dielectric layer of graphene/PDMS sponge sandwiched with patterned Cu antenna and electrode [41]; the air holes between the graphene particles (like numerous parallel mini-capacitors) are very sensitive to deformation. However, utilizing zero-dimensional nanoparticles or two-dimensional nanoflakes in the polymer matrix may likely cause the problem of aggregation and uneven dispersion. These shortcomings have prompted the pursuit of augmentative dielectric nanocomposites in the quest for high-performance pressure sensors. According to our understanding, there is no previous study based on porous polymer-based nanocomposite hybrids with one-dimensional nanostructures for capacitive pressure sensing applications.

Here, we propose a novel nanocomposite dielectric material where one-dimensional ZnO nanowire is incorporated into a porous elastomeric polymer for the formation of flexible capacitive pressure sensors. ZnO nanowire is a unique high-aspect-ratio, biocompatible, and low-cost material that exhibits semiconducting, piezoelectric, and pyroelectric multiple properties and that can be easily dispersed in water/organic solvent and polymer matrix. The porous polydimethylsiloxane (PDMS)-based nanocomposite of closed porosity is prepared by using a porogen-assisted process, and ZnO nanowires are randomly distributed in PDMS. Upon external compression, a large change in dielectric permittivity consolidating massive variations in capacitance can be achieved by enhancing the interfacial polarization and elastic modulus of the nanocomposite. The measured pressure sensitivity of the fabricated capacitive pressure sensors shows a remarkable improvement of more than 21 times when compared to that of flat PDMS devices; it can distinguish a subtle pressure of ~1.0 Pa with an ultrafine resolution as low as 0.4 Pa. The nanocomposite dielectric layer also reveals good stability and durability after 4000 loading–unloading cycles and a wide detection range, showing a great potential for the perception of various human motions.

2. Experimental

2.1. Growth of ZnO Nanowires

ZnO nanowires were produced by vapor phase carbothermic reduction [10,42,43]. A mixture of ZnO powder (99.9%, Alfa Aesar, Ward Hill, MA, USA) and graphite (Sigma-Aldrich, Burlington, MA, USA) (weight ratio ~1:1) in an alumina boat was loaded into the center of a quartz tube furnace, and a Si <111> substrate with a gold film (1.0 nm thick) was placed at the downstream end. Then, the temperature of the tube center was increased to 950 °C under an Ar/O_2 flow (50/1 sccm; total pressure: 2 mbar) to promote the nanowire growth.

2.2. Fabrication of ZnO Nanowire–Porous PDMS Capacitive Pressure Sensors

Prior to incorporation of nanowire into PDMS, as-grown ZnO nanowires on the Si substrate were dispersed into ethanol by sonication. The nanowires of different weight ratios (0, 0.5, 1.0, and 1.5 wt%, respectively) were blended with a proper amount of PDMS prepolymer (Sylgard® 184A, Dow Corning) by vigorous stirring to form a thorough viscous solution. After ethanol removing, the prepolymer blend was mixed with a porogen solution containing water/2-propanol (volume ratio of 3:1) at 3600 rpm for 30 min for the formation of porous elastomers [44]. The solution of pure PDMS or ZnO nanowire–PDMS prepolymers with porogen was further mixed with a curing agent (Sylgard® 184B) (weight ratio ~10:1) at 2000 rpm for 60 min in order to produce well-dispersed microdroplets in the PDMS-based solution.

Further, the proposed capacitors were fabricated on a transparent film of polyethylene terephthalate (PET, Nan Ya Plastics, thickness ~100 μm). Bottom electrodes of poly(3,4-ethylenedioxythiophene):poly(styrenesulfonate) (PEDOT:PSS, Clevios™ PH 1000) polymer (thickness: ~5 μm) were ink-jet printed by a commercial printer (Epson L120). Thus, the ZnO nanowire–PDMS prepolymer solution containing microdroplets of water could be blade-casted onto the PET with a subsequent baking at 70 °C for 24 h in order to achieve the thermal curing and solidification of the PDMS. In the meantime, the porogen microdroplets (water/2-propanol) confined in the PDMS could completely evaporate and permeate through PDMS at the curing temperature. Note that the water/2-propanol (volume ratio of 3:1) porogen provides an increased distribution of microdroplets inside PDMS, when compared with the porous network obtained using water alone as porogen.

Finally, a fully polymerized ZnO nanowire–porous PDMS film (thickness of ~200 μm) could be generated. The top PEDOT:PSS electrodes were deposited on the top of the dielectric layer, forming a sandwich-like capacitor layout (area: 10 mm × 10 mm). As a comparison, pristine porous PDMS (without nanowires) and flat PDMS (without pores and nanowires) were prepared at the same time via a similar manner. Note that the flat PDMS film (~200 μm) on a PEDOT:PSS/PET substrate was formed by casting its prepolymer, mixed with curing agent at a ratio of 10:1 (w/w), degassing in a vacuum oven for 30 min, and curing in the oven at 70 °C for 4–6 h.

2.3. Characterization and Measurement

The morphology and crystalline structure of the ZnO nanowires were characterized by high-resolution transmission electron microscopy (HRTEM, JEM-2100F CS STEM, JEOL, Akishima, Japan) and X-ray diffraction (D8 DISCOVER Plus-TXS, Bruker, Billerica, MA, USA). The cross-sectional morphology of flat PDMS, porous PDMS, and ZnO nanowire–porous PDMS nanocomposites was analyzed by scanning electron microscopy (SEM, SU8000 FE-SEM, HITACHI, Tegama, Japan). Capacitance vs. pressure loading measurements were measured on Agilent E4980AL Precision Impedance Analyze (at 860 kHz frequency with A.C. bias of 2.0 V). The relative dielectric permittivity was measured using a dielectric test fixture (Agilent 16451B, Keysight, Santa Rosa, CA, USA). The pressure sensing performance of the ZnO nanowire–porous PDMS was presented alongside those of the flat PDMS and pristine porous PDMS devices. All experiments were performed at room temperature.

3. Results and Discussion

A schematic flow for the fabrication of the proposed flexible capacitive pressure sensors based on a nanocomposite dielectric layer of ZnO nanowire and porous PDMS elastomer is illustrated in Figure 1. These pressure sensors were fabricated on PET substrates with PEDOT:PSS conducting electrodes (thickness: ~5 µm). Four porous composites were studied with 0, 0.5, 1.0, and 1.5 wt% of nanowires, respectively. The nanocomposite dielectric layer of ZnO nanowire–porous PDMS nanocomposite (thickness: ~200 µm) was formed by blade-casting. The incorporated wurtzite ZnO nanowires were ~30–70 nm in diameter and ~5–7 µm in length and revealed crystalline characteristics throughout by SEM, selected area electron diffraction (see Figure 2a), and X-ray diffraction. The HRTEM image displays (0002) lattice fringes with interplanar spacings of ~2.6 Å (Figure 2b). As can be seen from Figure 2c,d, both porous PDMS and ZnO nanowire–porous PDMS elastomers exhibit similar porous morphology, where the pore sizes were observed to be 4.2 ± 1.8 µm (see the Supplementary Information, Figure S1) with a calculated porosity of ~30.0%. Note that the porosity of the PDMS film can be adjusted through the mixing of porogen solution and PDMS polymer with different weight ratios. The cross-sectional SEM image of the nanocomposite films also proved that those nanowires (indicated by arrow marks) were distributed uniformly in the PDMS matrix without interrupting the porous networks. Further, the visual appearance of the nanocomposite film was clear and uniform; the ATR-FTIR (attenuated total reflection-Fourier transform infrared) spectra of porous PDMS and ZnO nanowire–porous PDMS were almost identical (Figure S2). Digital photographs of the fabricated capacitive pressure sensors are also shown in Figure 2e.

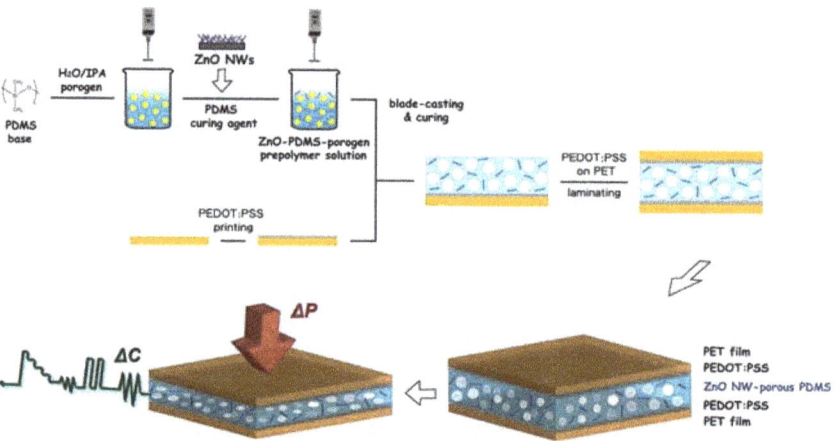

Figure 1. Schematic illustration for fabricating a porous polydimethylsiloxane-based nanocomposite dielectric film hybrid with zinc oxide nanowire for the formation of flexible capacitive pressure sensors.

Subsequently, we explored the pressure sensing capabilities of these porous capacitive pressure sensors with different loading of ZnO nanowires. To ensure the whole sensor was receiving uniform pressure load, a small plastic open container (contact area: ~10 mm × 10 mm, base pressure: ~10 Pa) was firmly attached to the sensor surface. A fixed amount of water droplets (20 µL) was carefully dispensed drop by drop into the container by using a micropipette (equivalent to a subtle pressure of ~2 Pa). The measured capacitance value of the sensor is denoted as C_0 (without pressure load) and C (with pressure load), respectively; the value can be determined based on $C = \varepsilon_r \varepsilon_0 A/d$, where ε_r and ε_0 are relative permittivity of the dielectric layer and permittivity of vacuum, respectively; d is the separation between two electrodes; and A is the area of the overlapped electrodes [45].

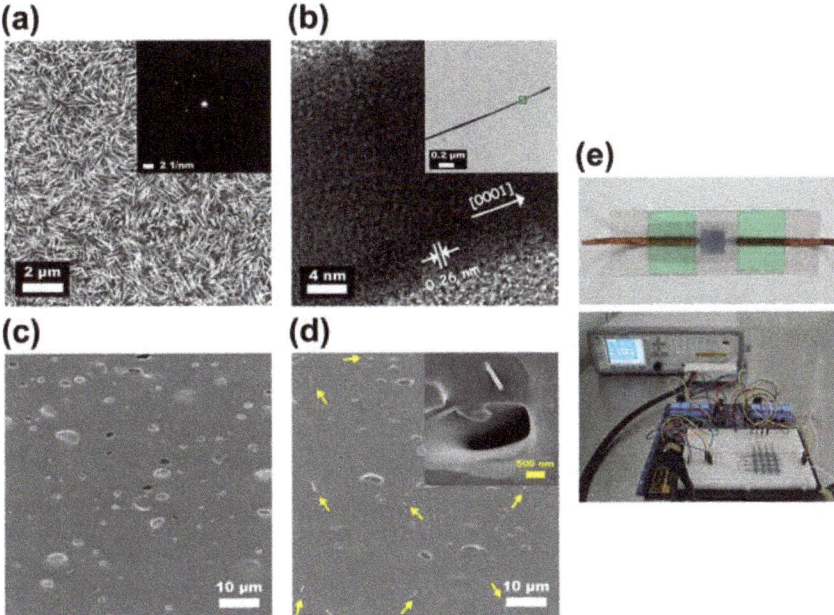

Figure 2. (**a**) SEM image of as-grown ZnO nanowires with ~30–70 nm diameter and ~5–7 μm length. Inset: a selected area electron diffraction pattern of a single-crystalline nanowire. (**b**) HRTEM images of a ZnO nanowire indicating the spacing of ~2.6 Å between two crystalline planes along [0001] growth direction. Cross-sectional SEM images of a portion of (**c**) a porous PDMS film and (**d**) a nanocomposite porous PDMS film with randomly distributed ZnO nanowires. Note that the arrow marks indicate the existence of nanowires. (**e**) Photo images of fabricated flexible ZnO nanowire–porous PDMS capacitive pressure sensors: a single cell (top) and a 4 × 4 multipixel array (bottom).

Figure 3a shows the relative capacitance change ($\Delta C/C_0$) as a function of the applied pressure (*P*) for the capacitive pressure sensors with different dielectric layers, where ΔC represents the change between C_0 and C. These curves, presenting how the capacitive pressure sensor senses the external pressure change, clearly undergo a sharp upward trend in the low-pressure regime and a smooth increment in the high-pressure regime. For flat PDMS sensors, the $\Delta C/C_0$ increased linearly with small applied pressure due to the elastic deformation between two electrodes. Further, under moderate-to-large pressure, the $\Delta C/C_0$ was nearly saturated, since the deformed PDMS became hard to deform. For porous PDMS sensors without nanowire loading, the sensor was compressed upon applied pressure, causing the embedded air pores to shrink. Thus, the sensor capacitance increased with the applied pressure under the combined effects of the pore thickness reduction and the dielectric permittivity increase, consequently promoting pressure sensitivity [36,44]. For ZnO nanowire–porous PDMS capacitive pressure sensors specifically, the performance in relative capacitance change was substantially better than those of the flat PDMS and porous PDMS devices. This was probably attributed to the elastic modulus of the nanocomposite dielectric elastomer, which can endow the dielectric film with good buffer effect upon external stress, and the gradual increase in dielectric permittivity during compressing (Figure S3). Consequently, the ZnO nanowire–porous PDMS elastomer can induce much higher variation in dielectric permittivity under the same compression pressure, which can operate properly over a wide range of 0–50 kPa.

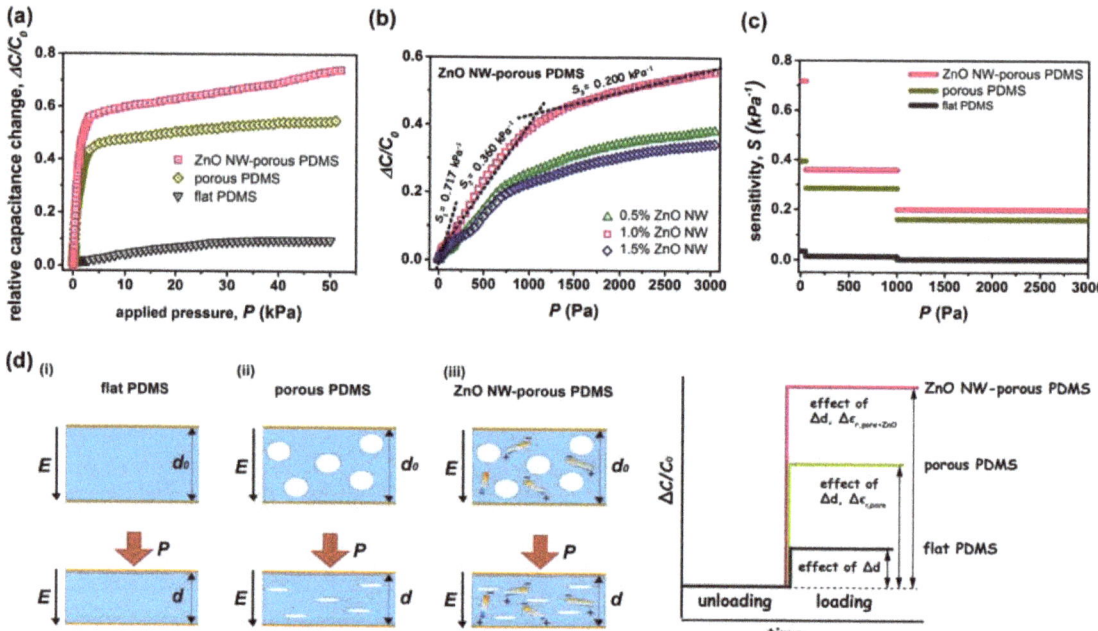

Figure 3. (**a**) Measured relative capacitance change ($\Delta C/C_0$) as a function of the applied pressure (P) for the capacitive pressure sensors with different types of dielectric layers: flat PDMS, porous PDMS, and ZnO nanowire–porous PDMS (sensing area: 10 mm × 10 mm). (**b**) Pressure-response plots for these nanocomposite sensors with varying ZnO nanowire loading. (**c**) Comparison of the sensitivity of capacitive pressure sensors at different applied pressure ranges. (**d**) Proposed sensing mechanisms with graphical capacitance change for the capacitors containing (i) flat PDMS, (ii) porous PDMS, and (iii) ZnO nanowire–porous PDMS.

By adjusting the nanowire loading, these ZnO nanowire–porous PDMS dielectric films present different pressure sensing capabilities. The relative capacitance changes of the ZnO nanowire (1.0 wt% loading)–porous PDMS nanocomposite were much more significant, which consequently gave the best comprehensive sensing performance (Figure 3b). This was closely related to the volumetric change of the porous structure and embedded nanowire of the nanocomposites during the compression process. The sensitivity of capacitive pressure sensors can be extracted through the slope of the curve of relative capacitive change versus applied pressure, as $S = \Delta (\Delta C/C_0)/\Delta P$ [45]. As can be seen, the ZnO nanowire (1.0 wt% loading)–porous PDMS sensor has the highest sensitivity of $S = 0.717$ kPa^{-1} in the regime of the 0–50 Pa range, presenting an improvement of 1.8 and 21.1 times over that of porous PDMS and flat PDMS, respectively. In the regime of 50–1000 Pa, the sensitivity of ZnO nanowire–porous PDMS, porous PDMS, and flat PDMS reduced to 0.360, 0.286, and 0.014 kPa^{-1}, respectively. In the regime of 1000–3000 Pa, notably, the ZnO nanowire–porous PDMS nanocomposite device still displayed an impressive pressure sensitivity of $S = 0.200$ kPa^{-1}, which was ~100 times greater than that of the flat PDMS sensor and ~1.4 times greater than that of the porous PDMS sensor. In contrast, the values of those porous PDMS and flat PDMS were degraded significantly because of their hardly compressed status. When the applied pressure was larger than 3000 Pa (up to 50 kPa in our test range), the pressure sensitivities of flat PDMS, porous PDMS, and ZnO nanowire–porous PDMS capacitors were <0.001 kPa^{-1}, ~0.002 kPa^{-1}, and ~0.004 kPa^{-1}, respectively. We also noticed that excessive content of ZnO nanowire loading (i.e., >1.0 wt%) may adversely affect the pressure response. For instance, the sensitivity of the porous PDMS-based device containing 1.5 wt% ZnO nanowires was reduced to

0.419 kPa^{-1}. This degradation clearly showed that excessive loading of nanowire did not positively contribute to the sensing performance, probably due to the increased elastic modulus of the nanocomposite. This can in turn diminish capacitance response and lead to a poorer sensitivity. Further, our attempts with a ZnO nanowire (2.0 wt% loading)–porous PDMS device did not succeed, since nonuniform nanowire distribution with certain aggregation was observed in the nanocomposite (Figure S4). This trade-off relationship is in line with previous observations in ceramic nanocrystals–PDMS capacitive pressure sensors [39]. Therefore, careful optimization and improvement on the interfacial compatibility of nanofillers and polymer matrix are necessary to maximize the positive contribution of ZnO nanowires. Overall, Figure 3c and Figure S5 summarize the measured maximum sensitivities and the mean values of relative capacitance change for the ZnO nanowire (1.0 wt% loading)–porous PDMS, porous PDMS, and flat PDMS capacitive pressure sensors at different applied pressure ranges.

To address the possible sensing mechanism, we anticipate that both air pores and ZnO nanowires incorporated in the elastomeric film provide beneficial effects when the proposed sensor is under compression. Figure 3d graphically depicts the relative capacitance change to the corresponding pressure loading for the sensors with flat PDMS, porous PDMS, and ZnO nanowire–porous PDMS dielectrics, respectively. As a parallel-plate capacitor, the variation of capacitance is determined by the dielectric property of the nanocomposite layer and the separation between two electrodes. Thus, under a uniform external pressure, the capacitance of the deformed flat PDMS film (see i) is simply a function of the reduced thickness of the dielectric film, since its relative permittivity ($\varepsilon_{r,flat}$) remains constant. Further, the porous PDMS (ii) can produce a larger deformation, since the presence of air pores in PDMS makes the pressure sensing material softer due to the reduced compressive modulus of the dielectric layer. These pore volumes can be gradually shrunk, and the air fraction in PDMS reduces; this can lead to an increase in the effective relative permittivity ($\Delta\varepsilon_{r,pore}$) and further enhance the change rate of capacitance as well as the pressure sensitivity. While the pressure loading is high, the pores will be nearly closed; the relative permittivity of the porous film will reach a saturation value, approaching the value of flat PDMS. Meanwhile, the mechanical property of the stressed porous PDMS film tends towards that of flat PDMS, becoming hardly compressible. For a porous nanocomposite–elastomer hybrid with nanowires (iii), the pressure sensing performance can be determined by comprehensive consideration of the elastic modulus of the dielectric and the change in relative permittivity (ε_r) during deformation. When the pressure loading is low, both the porous PDMS and ZnO nanowire–porous PDMS have similar air pore-induced deformation. This volumetric pore closing can in turn shorten the separation between ZnO nanowires and profoundly enhance the effective relative permittivity ($\Delta\varepsilon_{r,pore+ZnO}$) of the ZnO nanowire–porous PDMS layer, rather than that of porous PDMS. Thus, the total charge capacity and the change in capacitance will enhance intensely, leading to an augmented pressure sensing functionality. When the pressure loading is high, the pores in both films are nearly closed with very little ε_r increment. Thus, the sensitivity of the squeezed porous films is rather dominated by their elastic modulus. The flattened porous PDMS becomes hardly compressible; in contrast, the flattened ZnO nanowire–porous PDMS can still retain a certain deformability due to the nanowire-enhanced elastic modulus, contributing to a wider operation pressure range. In addition, we anticipate that the incorporation of ZnO nanowire can change the dielectric properties of the ZnO nanowire–porous PDMS nanocomposite. According to our measurement results, the capacitance (C_0) and relative permittivity (ε_r) of the nanocomposite film (C_0 = 9.80 pF, ε_r = 2.22 for 1.0 wt% nanowire loading) are higher than those of porous PDMS (8.94 pF, 2.02). This could prove that the addition of ZnO nanowire into porous PDMS matrix can enhance the dielectric polarization and relative permittivity due to the interfacial polarization between the ZnO nanowire and porous PDMS matrix. In short, the ZnO nanowire–porous PDMS has a remarkable sensing response compared to that of the pristine porous PDMS and flat PDMS, revealing much higher sensitivity under identical applied pressure.

Moreover, the time-resolved cyclic loading behaviors for a ZnO nanowire (1.0 wt% loading)–porous PDMS capacitive pressure sensor, a porous PDMS sensor, and a flat PDMS device are shown in Figure 4a. For the ZnO nanowire–porous PDMS nanocomposite device, a regular and steady sensing pattern was observed; the capacitance changed rapidly during pressure loading and unloading (applied pressure: 300 Pa). Its response time and recovery time were 260 ms and 400 ms, respectively. In contrast, the porous PDMS and flat PDMS devices both revealed slower response times of 300 ms and 650 ms as well as longer recovery retention times of 520 ms and 780 ms, respectively. Thereby, the incorporation of ZnO nanowires into the porous PDMS matrix can enhance the capacitance change and sensitivity profoundly. Moreover, to examine the limit of detection of the proposed sensor, water droplets of 10 mg per drop were carefully loaded by using a precision micropipette. As shown in Figure 4b, the additions of different droplets were clearly observed; specifically, the step increment of the capacitance change was primarily attributed to the applied water droplets, displaying a subtle pressure of only 1.0 Pa (one drop), 2.0 Pa (two drops), and 3.0 Pa (three drops), respectively. This demonstrates that our proposed sensor is rather responsive to external ultralow load. Meanwhile, we also placed different numbers of sesame seeds on the pressure sensor (Figure S6). The curve shows the progress from putting zero seeds to twelve seeds; the average seed weight is ~4 mg. Starting from three seeds being added, notably, the capacitance response presents a linear and steady increment each time a further seed is placed on the sensor, equating to an ultrafine resolution as low as only ~0.4 Pa. Hence, our device is quite sensitive and seemingly comparable with the limits of detection achieved in other studies: 0.1 Pa [36,37], 1 Pa [31,38], 5 Pa [41], 9 Pa [40], and 20 Pa [30].

In addition, we observed the long-term endurance of the ZnO nanowire–porous PDMS capacitive pressure sensor by allocating it to 4000 cycles of pressure loading–unloading (loading: 0.65 s; unloading: 0.65 s; applied pressure: 300 Pa). As depicted in Figure 4c and Figure S7 (for porous PDMS), the device kept very well in pressure sensing without obvious degradation during the test. For instance, mean signals under different cycles, i.e., between the sequence of 100−110, 1990−2000, and 3890−3900 (see insets), were almost the same (at loading: 10.79, 10.79, and 10.73 pF; at unloading: 9.80, 9.69, and 9.67 pF, respectively), presenting a negligible change of 0.6% in loaded capacitance and 1.3% in base capacitance. Therefore, our ZnO nanowire–porous PDMS sensor underwent negligible capacitance variation during the dynamic 4000 cyclic loading processes. It could be concluded that with the incorporation of ZnO nanowire the performance of the sensor is highly repeatable, durable, and stable.

Furthermore, we demonstrate that our proposed capacitive pressure sensor can maintain a reasonable response over a wide dynamic range, from 1.0 Pa to 50 kPa (by placing different objects, i.e., three sesame seeds, a paper clip, a playing card, a lighter, two game controllers, a standard weight, a speaker, a soft drink can, and even up to a ceramic mug). All in all, our nanocomposite pressure sensor can nearly cover the overall tactile pressure range. The recent progress on flexible capacitive pressure sensors, in terms of device components, sensitivity, limit of detection, response time, and operation pressure range, is also summarized in Table S1.

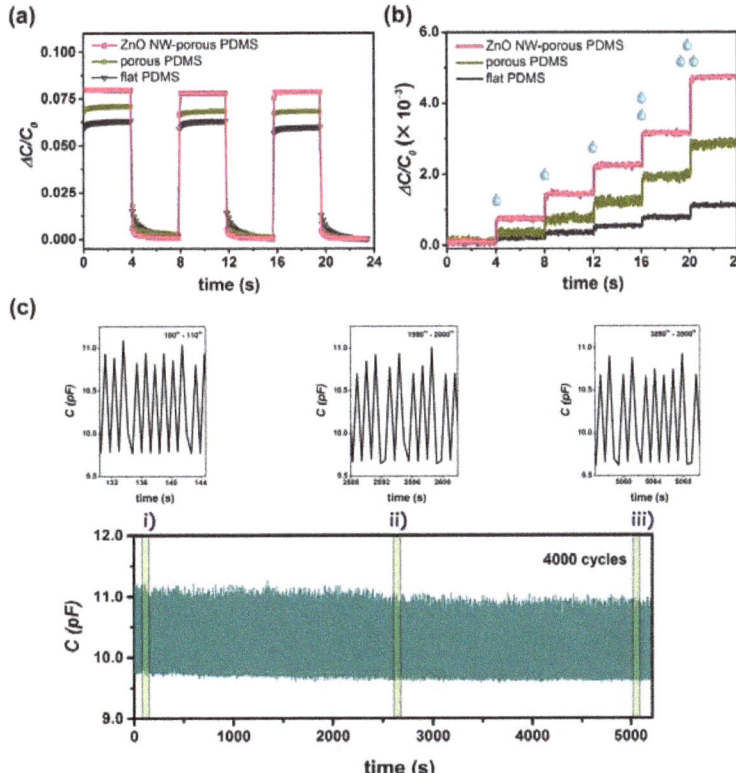

Figure 4. (a) Time-resolved capacitive response of nanocomposite capacitive pressure sensors based on flat PDMS, porous PDMS, and ZnO nanowire–porous PDMS, respectively. (b) Limit of detection test by means of the sequential detection of water droplets. (c) Stability test of a ZnO nanowire–porous PDMS device for 4000 cycles at 300 Pa; magnified curves of (i) 100−110, (ii) 1900−2000, and (iii) 3890−3900 cycles, respectively.

Based on the superior features of our ZnO nanowire–porous PDMS capacitive pressure sensor, the applications in several physiological activities were further investigated. As shown in Figure 5a, the sequential firm or soft touches given by a Chinese calligraphy apprentice can induce subtle pressure variations; on that account, our sensor can clearly detect and convert the force given to the brush into capacitances. In this case, the mean value of ΔC was ~0.95 pF, whereas the mean value for firm touches was ~1.2 pF and was ~0.65 pF for soft touches. Figure 5b shows the real-time capacitance response to an index finger excise. The finger bending and extending (in different bending angles of ~0°, 30°, 45°, 60°, or 90°) can change the stress applied onto the attached capacitor. When the finger was bent at different positions, the relative capacitance change increased accordingly; when the finger was fully straightened, the value then restored to its original state. Besides direct-contact stimuli, we also demonstrated the sensing of non-contact airflow by using an air spray gun. As can be seen in Figure 5c, the discernible signal peaks corresponding to the rapidly applied pulsed airflows can be observed simultaneously, which indicates the potency of adapting our proposed devices for non-contact signal monitoring. Another attempt presented in Figure 5d is for vocal cord movement detection, where a fabricated ZnO nanowire–porous PDMS sensor was placed at the throat region of a volunteer by using adhesive tape. Rapid and obvious responses with different patterns of signals were successfully recorded while speaking "one–two–three–four–five". Though not optimized yet, these demonstrations express the potential of utilizing our ZnO nanowire–porous PDMS sensor as a wearable device for human motion monitoring. In addition, a prototype 4 × 4 multipixel sensor array based on the proposed nanocomposite dielectric layer was fabricated for spatial pressure distribution tests. The area of each pixel was 5 mm × 5 mm, and the spacing between electrodes was 5 mm. As shown in Figure 5e, four different silicone rubber stamps (shaped with "N", "C", "T", and "U") were applied to the sensor array. Accordingly, each letter, which was presented in terms of induced capacitance change at the loaded pixels, was legible, and the crosstalk between pixels was rather small. Overall, the prototype sensor array can proceed with spatial pressure mapping, revealing potential for use in human–machine interface applications.

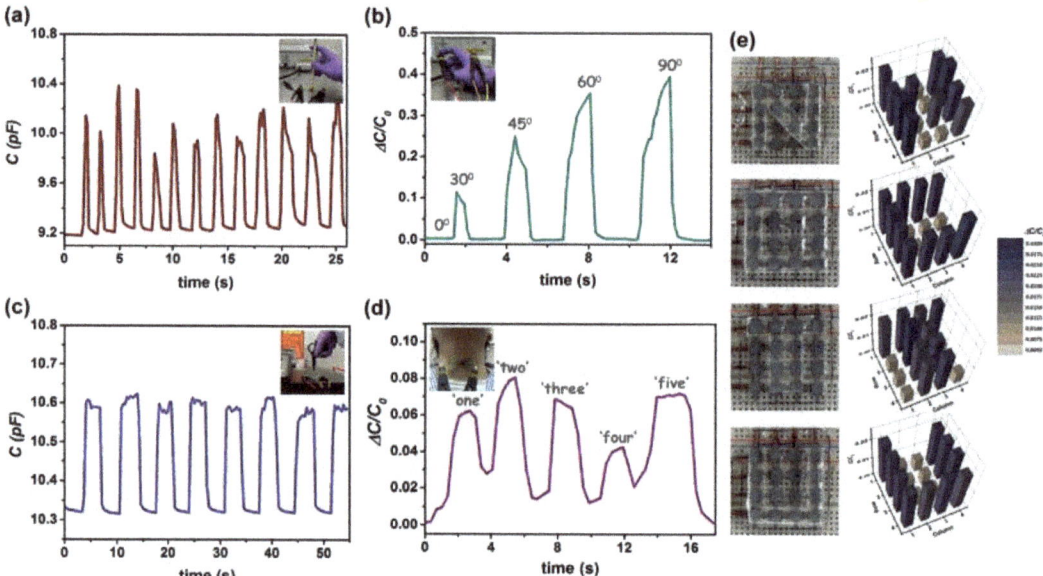

Figure 5. Real-time capacitance variations of the sensor in response to (**a**) calligraphy writing, (**b**) index finger straightening and bending, (**c**) air streaming, and (**d**) vocal speaking. (**e**) Photographs of a working pressure sensor array and their spatial pressure responses to the letters N, C, T, and U.

As a specific application demonstration, we have successfully applied our capacitive pressure sensor for water pressure monitoring. As shown in Figure S8, a sensor attached to a supporting glass slide was placed inside a vertical acrylic cylinder (length: 120 cm) filled with water. The device can clearly transduce underwater pressure at different underwater levels; the change in capacitance presents a sharp slope in the range of shallower water levels (0–5 cm) and a gentle slope in the range of deeper water levels (5–100 cm). By this experiment, the ZnO nanowire–porous PDMS pressure sensor is proved to be able to detect tiny changes in water pressure and, thus, may be helpful for underwater pressure sensing research.

In brief, this report displays the novelty of developing ZnO nanowire–porous PDMS nanocomposite dielectrics for wearable, wide-range, and low detection limit capacitive pressure sensors, which have been found to be highly responsive to subtle stimuli but vigorously so to gentle touch and verbal stimulation. For hybrid stimuli-sensitive nanostructure–porous polymer nanocomposites, there is plenty of potential for further enhancement. For instance, finding a novel way (perhaps by using hierarchical self-assembly, femtosecond laser, 3D printing, emulsion-templated phase separation, etc.) to control the size, distribution, and position of the pores in polymer matrix is intriguing; this can influence the sensor performance and reliability for practical applications. Experiments are also being conducted to examine chemical/physical stable materials (such as Ecoflex, polyurethane, polyimide, styrene-ethylene-butylene-styrene, etc.) and optimize device properties and to evaluate the working stability under different environmental conditions (e.g., temperature, humidity, underwater/oil). Additional discussion will center on the effect of interfacial affinity and interfacial polarization of functionalized nanostructures with various dielectric polymers. Overall, the demonstration of the present strategy may serve as a source of inspiration for the development of next-generation flexible pressure sensors.

4. Conclusions

To summarize, a novel nanocomposite capacitive pressure sensor was developed by hybridizing zinc oxide nanowire into a porous polydimethylsiloxane dielectric layer. By choosing the appropriate loading of nanowire into the matrix, the maximum sensitivity of the capacitive sensor was 0.717 kPa^{-1} at a subtle pressure range and 0.200 kPa^{-1} at a medium pressure range, exhibiting a broad detection range from as low as a sesame seed up to as high as a ceramic mug and revealing excellent operation stability over 4000 cycles. Furthermore, practical application measurements further showcased that this nanocomposite porous pressure sensor was capable of detecting finger bending, calligraphy writing, voice-induced vibration, air blowing, and hydrostatic underwater pressure, as well as identifying object spatial distribution. The results signify that these flexible nanowire–porous elastomeric nanocomposites could be used in capacitive-based pressure sensors for diverse applications, such as health monitoring and skin-like electronics.

Supplementary Materials: The following supporting information can be downloaded at: https://www.mdpi.com/article/10.3390/nano12020256/s1. Figure S1. Diameter distribution of generated air pores. Figure S2. ATR-FTIR spectra of porous films. Figure S3. Nanoindentation load-displacement curves. Figure S4. Cross-sectional SEM image of a ZnO nanowire (2 wt %)-porous PDMS film. Figure S5. Mean values of relative change in capacitance for capacitive pressure sensors at different pressure regimes. Figure S6. Plots of relative change in capacitance ($\Delta C/C_0$) versus applied pressure (P) by means of the sequential placement of sesame seeds. Figure S7. Operational stability test of the porous PDMS capacitive pressure sensor. Figure S8. Measured relative capacitance change as a function of hydrostatic water pressure. Table S1. Comparison of the ZnO nanowire-porous PDMS device with other recently reported capacitive pressure sensors.

Author Contributions: G.-W.H. designed and supervised the study. G.-W.H., L.-C.S., and P.-Y.C. conducted the experiments, data analysis, and interpretation. L.-C.S. and P.-Y.C. conducted characterization and pressure experiments. G.-W.H. drafted the manuscript with input from all coauthors. All authors have read and agreed to the published version of the manuscript.

Funding: This work was financially supported by the Ministry of Science and Technology, Taiwan (MOST 109-2221-E-009-061 & 110-2221-E-049-141) and the Higher Education Sprout Project of the National Yang Ming Chiao Tung University of Ministry of Education (MOE), Taiwan. The APC was funded by MOST 110-2221-E-049-141.

Institutional Review Board Statement: Not applicable.

Informed Consent Statement: Informed consent was obtained from all subjects involved in the study.

Data Availability Statement: Data is contained within the article and the supplementary materials.

Acknowledgments: The authors acknowledge the instrumental support of Taiwan Semiconductor Research Institute (JDP110-Y1-051).

Conflicts of Interest: There are no conflict to declare.

References

1. Zhao, S.; Zhu, R. Electronic skin with multifunction sensors based on thermosensation. *Adv. Mater.* **2017**, *29*, 1606151. [CrossRef]
2. Mazzotta, A.; Carlotti, M.; Mattoli, V. Conformable on-skin devices for thermo-electro-tactile stimulation: Materials, design, and fabrication. *Mater. Adv.* **2021**, *2*, 1787–1820. [CrossRef]

3. Chen, Z.; Zhao, D.; Ma, R.; Zhang, X.; Rao, J.; Yin, Y.; Wang, X.; Yi, F. Flexible temperature sensors based on carbon nanomaterials. *J. Mater. Chem. B* **2021**, *9*, 1941–1964. [CrossRef] [PubMed]
4. Xu, S.; Fan, Z.; Yang, S.; Zuo, X.; Guo, Y.; Chen, H.; Pan, L. Highly flexible, stretchable, and self-powered strain-temperature dual sensor based on free-standing PEDOT:PSS/carbon nanocoils–poly(vinyl) alcohol films. *ACS Sens.* **2021**, *6*, 1120–1128. [CrossRef] [PubMed]
5. Wu, Z.; Ding, H.; Tao, K.; Wei, Y.; Gui, X.; Shi, W.; Xie, X.; Wu, J. Ultrasensitive, stretchable, and fast-response temperature sensors. *ACS Appl. Mater. Interfaces* **2021**, *13*, 21854–21864. [CrossRef] [PubMed]
6. Gong, L.; Wang, X.; Zhang, D.; Ma, X.; Yu, S. Flexible wearable humidity sensor based on cerium oxide/graphitic carbon nitride nanocomposite self-powered by motion-driven alternator and its application for human physiological detection. *J. Mater. Chem. A* **2021**, *9*, 5619–5629. [CrossRef]
7. Riazi, H.; Taghizadeh, G.; Soroush, M. MXene-based nanocomposite sensors. *ACS Omega* **2021**, *6*, 11103–11112. [CrossRef]
8. Dai, G.; Wang, L.; Cheng, S.; Chen, Y.; Liu, X.; Deng, L.; Zhong, H. Perovskite Quantum dots based optical Fabry–Pérot pressure sensor. *ACS Photonics* **2020**, *7*, 2390–2394. [CrossRef]
9. Ravi, S.K.; Paul, N.; Suresh, L.; Salim, A.T.; Wu, T.; Wu, Z.; Jones, M.R.; Tan, S.C. Bio-photocapacitive tactile sensors as a touch-to-audio braille reader and solar capacitor. *Mater. Horiz.* **2020**, *7*, 866–876. [CrossRef]
10. Chen, Y.-S.; Hsieh, G.-W.; Chen, S.-P.; Tseng, P.-Y.; Wang, C.-W. Zinc oxide nanowire-poly(methyl methacrylate) dielectric layers for polymer capacitive pressure sensors. *ACS Appl. Mater. Interfaces* **2015**, *7*, 45–50. [CrossRef]
11. Claver, U.P.; Zhao, G. Recent progress in flexible pressure sensors based electronic skin. *Adv. Eng. Mater.* **2021**, *23*, 2001187. [CrossRef]
12. Lee, S.; Franklin, S.; Hassani, F.A.; Yokota, T.; Nayeem, M.O.G.; Wang, Y.; Leib, R.; Cheng, G.; Franklin, D.W.; Someya, T. Nanomesh pressure sensor for monitoring finger manipulation without sensory interference. *Science* **2020**, *370*, 966–970. [CrossRef]
13. Kim, D.W.; Kong, M.; Jeong, U. Interface design for stretchable electronic devices. *Adv. Sci.* **2021**, *8*, 2004170. [CrossRef] [PubMed]
14. Kim, K.-H.; Hong, S.K.; Jang, N.-S.; Ha, S.-H.; Lee, H.W.; Kim, J.-M. Wearable resistive pressure sensor based on highly flexible carbon composite conductors with irregular surface morphology. *ACS Appl. Mater. Interfaces* **2017**, *9*, 17499–17507. [CrossRef]
15. Kim, S.; Amjadi, M.; Lee, T.-I.; Jeong, Y.; Kwon, D.; Kim, M.S.; Kim, K.; Kim, T.-S.; Oh, Y.S.; Park, I. Wearable, ultrawide-range, and bending-insensitive pressure sensor based on carbon nanotube network-coated porous elastomer sponges for human interface and healthcare devices. *ACS Appl. Mater. Interfaces* **2019**, *11*, 23639–23648. [CrossRef] [PubMed]
16. Feng, P.; Zhong, M.; Zhao, W. Stretchable multifunctional dielectric nanocomposites based on polydimethylsiloxane mixed with metal nanoparticles. *Mater. Res. Express* **2020**, *7*, 015007. [CrossRef]
17. Zhang, Y.; Chi, Q.; Liu, L.; Zhang, T.; Zhang, C.; Chen, Q.; Wang, X.; Lei, Q. PVDF-based dielectric composite films with excellent energy storage performances by design of nanofibers composition gradient structure. *ACS Appl. Energy Mater.* **2018**, *1*, 6320–6329. [CrossRef]
18. Li, R.; Zhou, Q.; Bi, Y.; Cao, S.; Xia, X.; Yang, A.; Li, S.; Xiao, X. Research progress of flexible capacitive pressure sensor for sensitivity enhancement approaches, 321. *Sens. Actuators A* **2021**, *321*, 112425. [CrossRef]
19. Zhang, L.; Zhang, S.; Wang, C.; Zhou, Q.; Zhang, H.; Pan, G.-B. Highly sensitive capacitive flexible pressure sensor based on a high-permittivity MXene nanocomposite and 3D network electrode for wearable electronics. *ACS Sens.* **2021**, *6*, 2630–2641. [CrossRef]
20. Park, K.-I.; Jeong, C.K.; Kim, N.K.; Lee, K.J. Stretchable piezoelectric nanocomposite generator. *Nano Convergence* **2016**, *3*, 12. [CrossRef]
21. Nie, J.; Zhu, L.; Zhai, W.; Berbille, A.; Li, L.; Wang, Z.L. Flexible piezoelectric nanogenerators based on P(VDF-TrFE)/ CsPbBr$_3$ quantum dot composite films. *ACS Appl. Electron. Mater.* **2021**, *3*, 2136–2144. [CrossRef]
22. Zhai, W.; Nie, J.; Zhu, L. Enhanced flexible poly(vinylidene fluoride-trifluoroethylene) piezoelectric nanogenerators by SnSe nanosheet doping and solvent treatment. *ACS Appl. Mater. Interfaces* **2021**, *13*, 32278–32285. [CrossRef]
23. Chen, J.; Guo, H.; He, X.; Liu, G.; Xi, Y.; Shi, H.; Hu, C. Enhancing performance of triboelectric nanogenerator by filling high dielectric nanoparticles into sponge PDMS film. *ACS Appl. Mater. Interfaces* **2015**, *8*, 736–744. [CrossRef]
24. Kim, S.-R.; Yoo, J.-H.; Park, J.-W. Using electrospun AgNW/P(VDF-TrFE) composite nanofibers to create transparent and wearable single-electrode triboelectric nanogenerators for self-powered touch panels. *ACS Appl. Mater. Interfaces* **2019**, *11*, 15088–15096. [CrossRef] [PubMed]
25. Feng, P.-Y.; Xia, Z.; Sun, B.; Jing, X.; Li, H.; Tao, X.; Mi, H.-Y.; Liu, Y. Enhancing the performance of fabric-based triboelectric nanogenerators by structural and chemical modification. *ACS Appl. Mater. Interfaces* **2021**, *13*, 16916–16927. [CrossRef]
26. Zhang, F.; Zang, Y.; Huang, D.; Di, C.; Zhu, D. Flexible and self-powered temperature–pressure dual-parameter sensors using microstructure frame-supported organic thermoelectric materials. *Nature Comm.* **2015**, *6*, 8356. [CrossRef] [PubMed]
27. Sultana, A.; Ghosh, S.K.; Alam, M.M.; Sadhukhan, P.; Roy, K.; Xie, M.; Bowen, C.R.; Sarkar, S.; Das, S.; Middya, T.R.; et al. Methylammonium lead iodide incorporated poly(vinylidene fluoride) nanofibers for flexible piezoelectric–pyroelectric nanogenerator. *ACS Appl. Mater. Interfaces* **2019**, *11*, 27279–27287. [CrossRef]
28. Tian, Z.; Zhang, H.; Xiu, F.; Zhang, M.; Zou, J.; Ban, C.; Nie, Y.; Jiang, W.; Hub, B.; Liu, J. Wearable and washable light/thermal emitting textiles. *Nanoscale Adv.* **2021**, *3*, 2475–2480. [CrossRef]

29. Lee, S.; Kim, E.H.; Yu, S.; Kim, H.; Park, C.; Lee, S.W.; Han, H.; Jin, W.; Lee, K.; Lee, C.E.; et al. Polymer-laminated Ti$_3$C$_2$TX MXene electrodes for transparent and flexible field-driven electronics. *ACS Nano* **2021**, *15*, 8940–8952. [CrossRef]
30. Kang, B.-C.; Park, S.-J.; Ha, T.-J. Wearable pressure/touch sensors based on hybrid dielectric composites of zinc oxide nanowires/poly(dimethylsiloxane) and flexible electrodes of immobilized carbon nanotube random networks. *ACS Appl. Mater. Interfaces* **2021**, *13*, 42014–42023. [CrossRef] [PubMed]
31. Hsieh, G.-W.; Ling, S.-R.; Hung, F.-T.; Kao, P.-H.; Liu, J.-B. Enhanced piezocapacitive response in zinc oxide tetrapod–poly(dimethylsiloxane) composite dielectric layer for flexible and ultrasensitive pressure sensor. *Nanoscale* **2021**, *13*, 6076–6086. [CrossRef] [PubMed]
32. Lipomi, D.; Vosgueritchian, M.; Tee, B.C.-K.; Hellstrom, S.L.; Lee, J.A.; Fox, C.-H.; Bao, Z. Skin-like pressure and strain sensors based on transparent elastic films of carbon nanotubes. *Nat. Nanotechnol.* **2011**, *5*, 788–792. [CrossRef]
33. Viry, L.; Levi, A.; Totaro, M.; Mondini, A.; Mattoli, V.; Mazzolai, B.; Beccai, L. Flexible three-axial force sensor for soft and highly sensitive artificial touch. *Adv. Mater.* **2014**, *26*, 2659–2664. [CrossRef] [PubMed]
34. Mannsfeld, C.B.S.; Tee, B.C.-K.; Stoltenberg, R.M.; Chen, C.V.H.-H.; Barman, S.; Muir, B.V.O.; Sokolov, A.N.; Reese, C.; Bao, Z. Highly sensitive flexible pressure sensors with microstructured rubber dielectric layers. *Nat. Mater.* **2010**, *9*, 859–864. [CrossRef]
35. Chen, S.; Zhuo, B.; Guo, X. Large area one-step facile processing of microstructured elastomeric dielectric film for high sensitivity and durable sensing over wide pressure range. *ACS Appl. Mater. Interfaces* **2016**, *8*, 20364–20370. [CrossRef]
36. Kwon, D.; Lee, T.-I.; Shim, J.; Ryu, S.; Kim, M.S.; Kim, S.; Kim, T.-S.; Park, I. Highly sensitive, flexible, and wearable pressure sensor based on a giant piezocapacitive effect of three-dimensional microporous elastomeric layer. *ACS Appl. Mater. Interfaces* **2016**, *8*, 16922–16931. [CrossRef] [PubMed]
37. Yang, J.C.; Kim, J.-O.; Oh, J.; Kwon, S.Y.; Sim, J.Y.; Kim, D.W.; Choi, H.B.; Park, S. Microstructured porous pyramid-based ultrahigh sensitive pressure sensor insensitive to strain and temperature. *ACS Appl. Mater. Interfaces* **2019**, *11*, 19472–19480. [CrossRef]
38. Luo, Y.; Shao, J.; Chen, S.; Chen, C.; Tian, H.; Li, X.; Wang, L.; Wang, D.; Lu, B. Flexible capacitive pressure sensor enhanced by tilted micropillar arrays. *ACS Appl. Mater. Interfaces* **2019**, *11*, 17796–17803. [CrossRef]
39. Mu, C.; Li, J.; Song, Y.; Huang, W.; Ran, A.; Deng, K.; Huang, J.; Xie, W.; Sun, R.; Zhang, H. Enhanced piezocapacitive effect in CaCu$_3$Ti$_4$O$_{12}$–polydimethylsiloxane composited sponge for ultrasensitive flexible capacitive sensor. *ACS Appl. Nano Mater.* **2018**, *1*, 274–283. [CrossRef]
40. Pruvost, M.; Smit, W.J.; Monteux, C.; Poulin, P.; Colin, A. Polymeric foams for flexible and highly sensitive low-pressure capacitive sensors. *Npj Flex. Electron.* **2019**, *3*, 7. [CrossRef]
41. Kou, H.; Zhang, L.; Tan, Q.; Liu, G.; Dong, H.; Zhang, W.; Xiong, J. Wireless wide-range pressure sensor based on graphene/PDMS sponge for tactile monitoring. *Sci. Rep.* **2019**, *9*, 3916. [CrossRef] [PubMed]
42. Li, F.M.; Hsieh, G.-W.; Dalal, S.; Newton, M.; Stott, J.E.; Hiralal, P.; Nathan, A.; Warburton, P.A.; Unalan, H.E.; Beecher, P.; et al. Zinc oxide nanostructures and high electron mobility nanocomposite thin film transistors. *IEEE Trans. Electron. Dev.* **2008**, *55*, 3001–3011. [CrossRef]
43. Chen, S.-P.; Chen, Y.-S.; Hsieh, G.-W. N-Channel zinc oxide nanowire: Perylene diimide blend organic thin film transistors. *IEEE J. Electron. Dev. Soc.* **2017**, *5*, 367–371. [CrossRef]
44. Kwak, Y.; Kang, Y.; Park, W.; Jo, E.; Kim, J. Fabrication of fine-pored polydimethylsiloxane using an isopropyl alcohol and water mixture for adjustable mechanical, optical, and thermal properties. *RSC Adv.* **2021**, *11*, 18061–18067. [CrossRef]
45. Hammock, M.L.; Chortos, A.; Tee, B.C.-H.; Tok, J.B.-H.; Bao, Z. 25th anniversary article: The evolution of electronic skin (e-skin): A brief history, design considerations, and recent progress. *Adv. Mater.* **2013**, *25*, 5997–6037. [CrossRef]

Article

Triboelectric Response of Electrospun Stratified PVDF and PA Structures

Pavel Tofel [1,2], Klára Částková [2,3], David Říha [1,3], Dinara Sobola [1,4,5], Nikola Papež [1], Jaroslav Kaštyl [2,3], Ștefan Țălu [6,*] and Zdeněk Hadaš [7]

1. Department of Physics, Faculty of Electrical Engineering and Communication, Brno University of Technology, Technická 2848/8, 616 00 Brno, Czech Republic; tofel@vut.cz (P.T.); xrihad01@vutbr.cz (D.Ř.); sobola@vut.cz (D.S.); papez@vut.cz (N.P.)
2. Central European Institute of Technology, Purkyňova 656/123, 612 00 Brno, Czech Republic; klara.castkova@ceitec.vutbr.cz (K.Č.); jaroslav.kastyl@ceitec.vutbr.cz (J.K.)
3. Department of Ceramics and Polymers, Faculty of Mechanical Engineering, Brno University of Technology, Technická 2896/2, 616 69 Brno, Czech Republic
4. Institute of Physics of Materials, Czech Academy of Sciences, Žižkova 22, 616 62 Brno, Czech Republic
5. Department of Inorganic Chemistry and Chemical Ecology, Dagestan State University, St. M. Gadjieva 43-a, 367015 Makhachkala, Russia
6. Directorate of Research, Development and Innovation Management (DMCDI), Technical University of Cluj-Napoca, Constantin Daicoviciu Street, No. 15, 400020 Cluj-Napoca, Romania
7. Institute of Solid Mechanics, Mechatronics and Biomechanics, Faculty of Mechanical Engineering, Brno University of Technology, Technická 2896/2, 616 69 Brno, Czech Republic; Zdenek.Hadas@vut.cz
* Correspondence: stefan_ta@yahoo.com or stefan.talu@auto.utcluj.ro; Tel.: +40-264-401-200; Fax: +40-264-592-055

Abstract: Utilizing the triboelectric effect of the fibrous structure, a very low cost and straightforward sensor or an energy harvester can be obtained. A device of this kind can be flexible and, moreover, it can exhibit a better output performance than a device based on the piezoelectric effect. This study is concerned with comparing the properties of triboelectric devices prepared from polyvinylidene fluoride (PVDF) fibers, polyamide 6 (PA) fibers, and fibrous structures consisting of a combination of these two materials. Four types of fibrous structures were prepared, and then their potential for use in triboelectric devices was tested. Namely, individual fibrous mats of (*i*) PVDF and (*ii*) PA fibers, and their combination—(*iii*) PVDF and PA fibers intertwined together. Finally, the fourth kind was (*iv*), a stratified three-layer structure, where the middle layer from PVDF and PA intertwined fibers was covered by PVDF fibrous layer on one side and by PA fibrous layer on the opposite side. Dielectric properties were examined and the triboelectric response was investigated in a simple triboelectric nanogenerator (TENG) of individual or combined (*i–iv*) fibrous structures. The highest triboelectric output voltage was observed for the stratified three-layer structure (the structure of *iv* type) consisting of PVDF and PA individual and intertwined fibrous layers. This TENG generated 3.5 V at peak of amplitude at 6 Hz of excitation frequency and was most sensitive at the excitation signal. The second highest triboelectric response was observed for the individual PVDF fibrous mat, generating 2.8 V at peak at the same excitation frequency. The uniqueness of this work lies in the dielectric and triboelectric evaluation of the fibrous structures, where the materials PA and PVDF were electrospun simultaneously with two needles and thus created a fibrous composite. The structures showed a more effective triboelectric response compared to the fibrous structure electrospun by one needle.

Keywords: dielectric properties; electrospinning; fiber composite; PVDF; PA; TENG; triboelectric effect

1. Introduction

Energy harvesters and sensors based on the triboelectric effect have been a very promising research field in the last ten years. Their advantage is in a wide range of materials that can be used and the simple fabrication process of these triboelectric devices.

Many researchers work on different kinds of devices using the triboelectric effect. This effect can be used in devices such as self-powered textiles [1], wearable electronics [2], self-powered human motion sensors [3], self-powered automobile sensors [4], position sensing and touchpads [5], flexible nano generators [6], non-invasive biomedical monitoring systems [7], energy harvesters for devices in internet of things infrastructure [8], environmental monitoring systems [9], air filters [10], and topically very important research into protection against a coronavirus pandemic, where the simple triboelectric nanogenerator with an electrocution layer may serve the purpose of filtration and the deactivation of SARS-CoV-2 [11]. These perspectives lead us to study this phenomenon and utilize its potential in simple devices based on a fibrous structure. From a triboelectric series, a combination of PVDF (polyvinylidene fluoride) and PA (polyamide 6) creates a very effective triboelectric device [12].

The triboelectric effect and level of contact electrification performance are based on materials that come into contact. In this case, contact electrification occurs caused by a surface transference of electrons or ions between these two materials [13]. Each material has its own ability to lose or gain electrons during the contact electrification process. This ability can be found in a list of materials (triboelectric series) where polarity and amount of charge for each material are described. An example of the triboelectric series can be seen in Figure 1a (based on various works [14–17]). PVDF is a material which tends to gain electrons during electrification. On the other hand, PA is positively charged during electrification. Both materials lie far enough apart in the triboelectric series, and this distance indicates their high triboelectric potential on contact.

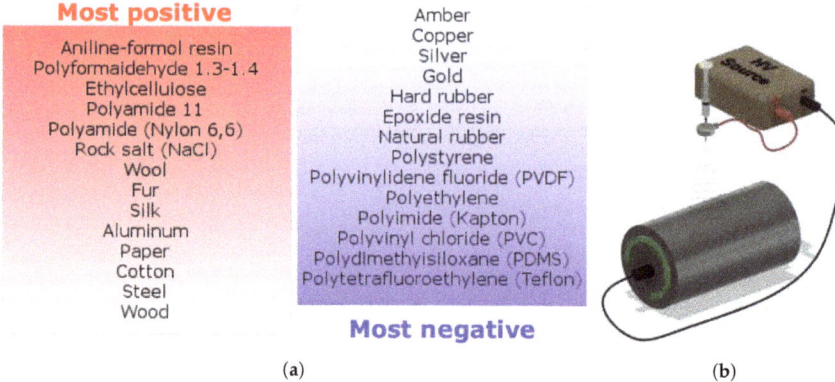

Figure 1. (a) A triboelectric series assembled from various works. (b) Scheme of electrospinning process.

One of the most used fabrication techniques for the preparation of fibrous structures, which can be used as tribomaterials, is electrospinning [18–20]. During the electrospinning, a polymer solution is ejected from a needle tip by applying a high voltage between the needle and a grounded collector, as illustrated in Figure 1b. When an electrostatic force overcomes the surface tension at the tip of the needle, a so-called Taylor's cone is formed, and it is elongated into a fluid jet. The jetting fluid is collected on a collector in the form of fibers due to the potential difference between the needle tip and the collector. The electrospinning process produces material in the form of long and thin nanofibers [18,21]. The electrospun material has a rough surface and a significantly larger surface area per unit volume than the material prepared as a bulk [22], which can be highly beneficial in applications such as triboelectric devices. Fibers significantly improve the triboelectric effect of electrospun materials, as their large surface area allows the generation of a large number of charges during electrification [23–25].

Electrospinning is often used to prepare materials where it is necessary to increase their surface area [26,27]. Mostly are reported the combinations of electrospun nanofibrous mats as one triboelectric layer against another triboelectric layer which is in solid-state [28], porous form [25], or in the form of some nanostructure [29]. All these modifications lead to enhancing the triboelectric output performance of the triboelectric device. The PVDF can be prepared in polarized and unpolarized forms, but no significant difference was reported between these two forms if the PVDF is used as a triboelectric device [23,26].

It is very interesting to study multilayer triboelectric devices, where an interlayer is situated between tribomaterial and electrode. This interlayer is used for charge-trapping and significantly increases the triboelectric output [26]. It has been demonstrated that triboelectric device assembled from PVDF cast on polyimide increased the triboelectric output 8× [30]. If the polydimethylsiloxane (PDMS) is used instead of polyimide, the triboelectric output can increase even more [31].

Our research into the triboelectric devices was focused on electrospun materials such as PVDF and PA, and their combinations in fiber form with the effort to evaluate their possibilities and performance in triboelectric devices. These materials are widely used in triboelectric devices for their non-reactivity, mechanical robustness, and flexibility [32]. Different combinations of tribomaterials were fabricated from these nanofiber materials, which were consequently tested from the triboelectric point of view. Only a few researchers are focused on the triboelectric properties of the fiber materials where two materials are intertwined between each other. For example, Garcia et al. present work where two fiber materials were sandwiched between electrodes, and this device can be used as a simple self-powered pressure sensor [33]. The fibrous materials in this work were prepared separately and then compressed between two electrodes. The PVDF and PA fibrous materials were electrospun individually or simultaneously using a two-needle setting. In order to study the triboelectric performance, the fibrous materials were arranged in different types of simple or stratified devices and characterized.

The V-Q-x Relationship for Contact-Separation Mode

The triboelectric nanogenerator (TENG) generally operates in one of four basic modes such as contact-separation (CS) mode, lateral sliding (LS) mode, single-electrode (SE) mode, and freestanding triboelectric-layer (FT) mode [34]. We have focused on CS mode because TENG is simple to manufacture and is very widely used. TENGs are commonly described by the V-Q-x relationship, where V represents the voltage between electrodes, Q is the amount of charge transferred between the electrodes, and x is a separation distance of the electrodes. The theoretical model for the contact separation mode is shown in Figure 2. Figure 2a represents TENG, where two dielectric materials are used, while Figure 2b represents TENG with one dielectric material.

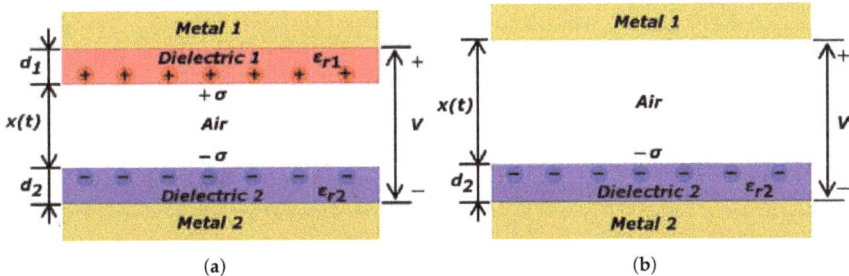

Figure 2. Contact-separation mode theoretical model. (**a**) Dielectric-to-dielectric and (**b**) conductor-to-dielectric.

The output voltage of TENG in contact mode can be expressed by V-Q-x relationship [35–37]:

$$V = -\frac{Q}{S\varepsilon_0}(d_0 + x(t)) + \frac{\sigma x(t)}{\varepsilon_0}, \qquad (1)$$

where Q is the charge, S is the area, ε_0 is the permittivity of space, $x(t)$ is the displacement of electrodes as a function of time, σ is the triboelectric charge density, and d_0 is given:

$$d_0 = \frac{d_1}{\varepsilon_{r1}} + \frac{d_2}{\varepsilon_{r2}}. \tag{2}$$

At open-circuit, the conditions are no charge transferred, and Q is equal to zero. Then open-circuit voltage V_{oc} is given:

$$V_{oc} = \frac{\sigma x(t)}{\varepsilon_0}. \tag{3}$$

On the contrary, at short circuit conditions, V is equal to zero. Therefore, a short circuit current I_{sc} is given:

$$I_{sc} = \frac{S\sigma d_0 v(t)}{(d_0 + x(t))^2}. \tag{4}$$

V_{oc} and I_{sc} are dependent on the triboelectric charge density σ. The charge density σ and space distance x influences V_{oc}, whereas the I_{sc} is further dependent on the contact speed [35–37]. Thus it can be concluded that the materials in the triboelectric devices can be compared in terms of V_{oc} values, where the speed of contact is ignored.

2. Material and Methods

The electrospun polyvinylidene fluoride (molar mass 275,000 g/mol, Sigma Aldrich, St. Louis, MO, USA) and polyamide 6 (molar mass 35,000 g/mol, Alfa Chemicals, Bracknell, UK) were used for electrospinning. Dimethylsulfoxide p.a. (DMSO, Sigma Aldrich), acetone p.a. (Ac, Sigma Aldrich), acetic acid (AA, Penta, Bratislava, Slovakia), and formic acid (FA, Merck, Darmstadt, Germany) were used for polymers solutions preparation [18–20,38,39]. For PVDF solutions, solvents DMSO and Ac were mixed in a volume ratio 7:3. The PVDF beads were dissolved in the binary solvent in a concentration of 20 wt% at 50 °C for 24 h until a visually homogeneous solution was formed. For PA solutions, solvents AA and FA were mixed in a volume ratio 8:2. The PA granules were dissolved in the binary mixture in a concentration of 15 wt% at lab temperature for 24 h until a clear solution was achieved.

The prepared solutions were electrospun using the 4SPIN electrospinning equipment (Contipro, Czech Republic) at a feeding rate of (10–20) µL/min through a needle. Needle diameter and rotation speed for each structure are shown in Table 1. The accelerating voltage was 40 kV, and the distance between the needle tip and the collector (rotating cylinder covered by aluminum foil) was 15 cm. These electrospinning parameters were constant for all structures. The fibrous samples PVDF and PA were collected for 90 min in the form of non-woven structures, which were further characterized.

A special arrangement was used for samples PVDF+PA and S(PVDF+PA). The combined spinning (co-spinning) using two independent needles with separate feeding by PVDF and PA was applied for 90 min to spin sample PVDF+PA with intertwining fibers. In the case of S(PVDF+PA) sample, the structure was spun layer by layer. Firstly, the PVDF layer was spun for 45 min, then PVDF+PA layer co-spun for 45 min and, finally, the PA layer was spun for 45 min. See Table 1 for the details of samples and processing parameters.

Table 1. Overview of samples and processing parameters.

Sample	Structure	Needle (–)	Rotation Speed (rpm)
PA	PA fibers	19G	300
PVDF	PVDF fibers	17G	300
PVDF+PA	PVDF+PA intertwined fibers	17G/19G	300
S(PVDF+PA)	PVDF//PVDF+PA//PA *	see above	2000

* Please note that the double slash used here is used to distinguish individual layers in the stacked structure.

The final thickness of our samples was measured by interferometer ILD 1402-10 (Micro Epsilon, Ortenburg, Germany). A thin metal plate in the shape of a square with 30 mm edge and a thickness of 0.1 mm was placed on the final fibrous structure, which was still on the Al foil. The thickness was then measured using an ILD 1402-10 at the center of the plate. After subtracting the thickness of the Al foil and the metal plate, the thickness of the fibrous structure was obtained.

Dielectric properties were measured by Alpha-A High Performance Modular Measurement System (Novocontrol, Montabaur, Germany). As a sample holder, the 16451B dielectric test fixture (Agilent, Tokyo, Japan) with a dimension of the active electrode 5 mm was used. The triboelectric energy performance of the prepared samples was evaluated by electrometer 6517b (Keithley, Solon, OH, USA). The TENG was assembled in vertical contact-separation mode. The moving part consisting of the Cu electrode was controlled by the vibration test system TV 50018 (Tira, Schalkau, Germany). The sample was clamped on a fixed Cu electrode. The area of the active part of the generator was (30×30) mm. Mechanical force was measured by force sensor 208C01 (PCB Piezotronics, Hückelhoven, Germany), and this sensor was situated on the side of the fixed electrode. The displacement between electrodes was measured via interferometer ILD 1402-10 (Micro Epsilon, Ortenburg, Germany). Output voltages generated by the triboelectric devices were measured by oscilloscope DSOX2024A (Keysight, Santa Rosa, CA, USA). The same oscilloscope was used for the evaluation of distance between electrodes measured by interferometer.

Dynamic Signal Acquisition Module NI-9234 (National Instruments, Austin, TX, USA) was used for the force sensor, and the communication was performed via SignalExpress software. Measured data were processed by Matlab R2018a software.

The triboelectric materials were arranged into simple triboelectric devices operating in contact-separation mode. Five triboelectric devices were assembled where four devices were the type of conductor-to-dielectric, and one triboelectric device was the dielectric-to-dielectric type, as is shown in Figure 3a. The measurement setup is shown in Figure 3b, which also shows the pressed and released state of the triboelectric device during the measurement. The area of the triboelectric material in TENG was (30×30) mm. Device A used PA fibers against copper electrode, device B used PVDF fibers against copper electrode, device C was based on the triboelectric layer of PVDF+PA sample with intertwined fibers, and device D was compiled from the triple-layer dielectric fibrous structures. Device E was assembled with PVDF fibers on one electrode against PA fibers on the other electrode, creating a triboelectric device where the two dielectrics come into contact during operation. This device is presented here as an example of the standard use of these two materials in the TENG device to exploit their maximum triboelectric potential. We can compare how triboelectric devices with one dielectric against a metal electrode perform compared to this standard device E, where two dielectric materials come into contact.

An upper electrode was fixed, and the bottom electrode was controlled by a shaker. For this simple comparison, the devices were connected to an oscilloscope over 10 MΩ probe.

The quasi-static piezoelectric constant d_{33} was measured using a Berlincourt d_{33} meter (YE2730A, Sinocera, China) on electrospun structures sandwiched between two copper electrodes. All the prepared fibrous structures exhibited zero or negligible piezoelectricity with a maximum value of up to 1 pC/N.

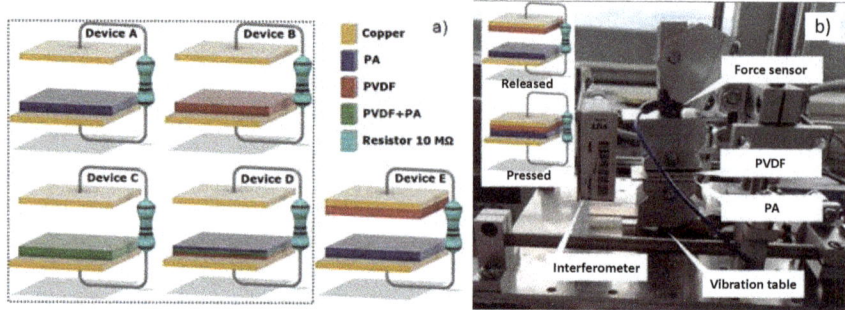

Figure 3. (a) Triboelectric devices assembled from different fibrous mats containing PVDF and PA material. The dashed line indicates triboelectric devices consisting of a single dielectric. (b) The measuring setup for triboelectric response measurement, where the inset figure illustrate pressed and released state of TENG during measurement.

3. Results and Discussion

Four fibrous structures were prepared by electrospinning, and the resulting SEM images of individual samples are shown in Figure 4. The PA sample (Figure 4a) consisted of slim fibers with a diameter around 100 nm. On the contrary, the thick fibers can be observed for the PVDF sample (Figure 4b), where the fiber diameter was approx. 1000 nm. A combination of slim and thick fibers can be observed in Figure 4c, where the structure of PVDF+PA sample is depicted. The structure of the layered S(PVDF+PA) sample is shown in Figure 4d and Figure 4e, where the top and bottom sides are shown, respectively. The cross-section of the layered S(PVDF+PA) sample is shown in Figure 4f.

3.1. Dielectric Properties of Electrospun Samples

Materials prepared in a fibrous form basically exhibit lower dielectric constant against their bulky counterparts due to the high porosity content [40]. Measurement of the dielectric properties is challenging for the fiber materials because the measurement strongly depends on the air involved in the fiber mats structure. Nevertheless, if the pressure during measurement is of the same value for all measured samples, then the dielectric constant and other dielectric properties can be compared with each other.

The dependence of the dielectric constant on frequency determined for prepared fibrous structures is shown in Figure 5a. The dielectric constant decreases with increasing frequency without any significant extremes for our samples, indicating our structures' dielectric behavior. It is necessary to note that all fibrous materials were prepared at a final thickness of around 20 μm.

Generally, the PA has the dielectric constant of around 4 in the dense form [37,41]. Our PA sample exhibited the dielectric constant $\varepsilon_r = 1.19$ at 1 kHz. The dielectric constant decreases almost linearly with increasing frequency. More significant reduction in dielectric constant was achieved for the sample formed by PVDF fibers. We have obtained the dielectric constant $\varepsilon_r = 1.57$ at 1 kHz for the PVDF sample, which is roughly one-tenth of pure and dense PVDF material with α-phase (the β-phase slightly reduces this value) [16]. This is due to the high porosity of the sample, as it is made up of thick sparse fibers, as can be seen from Figure 4b. The combination of PVDF and PA fibers created a structure where the dielectric constant is around $\varepsilon_r = 1.24$ at 1 kHz. The density of PVDF fibers is not large compared to PA fibers in the sample structure (see Figure 4c). Therefore, the dielectric constant's dependence is more similar to that measured on PA fibers. The last layered sample S(PVDF+PA) further increased the dielectric constant on $\varepsilon_r = 1.45$ at 1 kHz. The mentioned increase in the dielectric constant was due to the higher number of PVDF fibers in the S(PVDF+PA) sample compared to the PVDF+PA sample.

The dielectric losses versus frequency measured for the fabricated fibrous structures are shown in Figure 5b. The PA sample and the combination of PVDF and PA fibers (sample PVDF+PA) showed very low losses over the entire measured frequency spectrum. Both dependencies had a similar pattern and exhibited a dielectric loss of 0.7% at 1 kHz. Low losses were also observed at 1 Hz, where PA sample and the PVDF+PA sample exhibited dielectric loss 2.3% and 1.8%, respectively. On the other hand, the PVDF sample also exhibited low losses at 1 kHz (tan δ = 0.7%), but also high losses at low frequencies (tan δ = 6% at 1 Hz). Specified dielectric loss behavior is typical of fibrous PVDF materials, as can be seen in many works [41,42]. The layered S(PVDF+PA) sample showed the highest losses of 1.1% at 1% from all structures. It also exhibited high losses at low frequencies, where the dielectric loss was 6.6% at 1 Hz. The loss profile for this sample was similar to the PVDF fibrous sample, except that it was smoother.

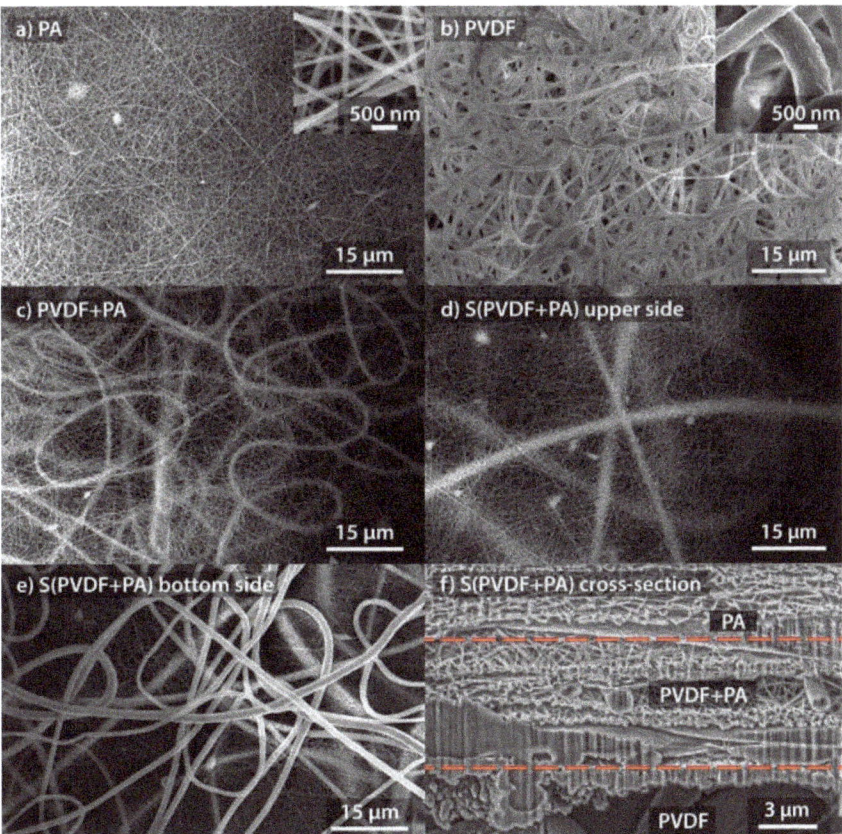

Figure 4. SEM images of (**a**) PA sample, (**b**) PVDF sample, (**c**) PVDF+PA sample, (**d**) top side of S(PVDF+PA) sample, (**e**) bottom side of S(PVDF+PA) sample (the side in contact with alumina foil during electrospinning process), and (**f**) cross-section of S(PVDF+PA) sample.

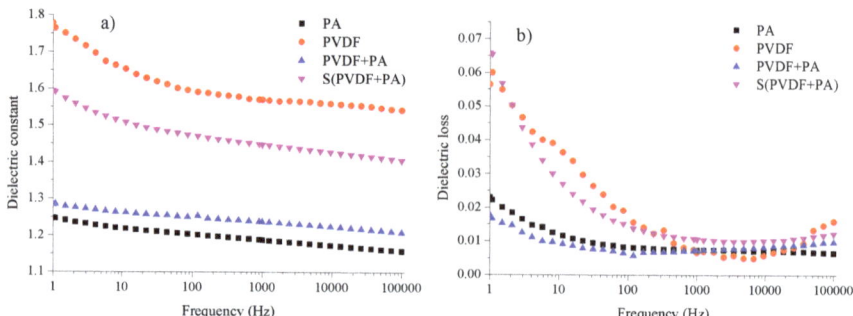

Figure 5. Dielectric properties of the prepared samples, (**a**) dielectric constant, and (**b**) dielectric loss in frequency dependence.

3.2. Triboelectric Properties of Electrospun Samples

The fibrous structures were measured in a simple TENG operating in the CS mode. In the case of the TENG device, we monitored the output voltage of the TENG when the electrode spacing and the electrode contact force were also observed. An example of such an investigation is shown in Figure 6b, where device D was measured. The output voltage of TENG was measured by an oscilloscope with the internal resistance of 10 MΩ, as can be seen in Figure 6a. The aim was to see how these TENGs perform if they were used as simple dynamic force sensors. And to evaluate, if the fibrous composites provide any benefit in these applications.

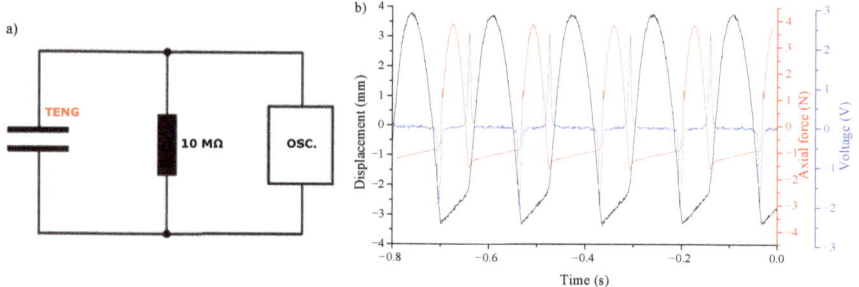

Figure 6. (**a**) Schematic of the measuring circuit. (**b**) Time record of the measurements on device D, where the output voltage, electrode displacement, and electrode contact force were recorded.

Assembled TENG devices were tested at different excitation frequencies, where the force (approx. 10 N) and maximal displacement (approx. 9 mm) between electrodes were constant. The excitation frequency range was applied from 2 Hz up to 10 Hz. The output voltages for individual TENGs formed by a single dielectric (conductor-to-dielectric CS TENG), for different excitation frequencies, are shown in Figure 7, where (a) is for the TENG labeled as device A, (b) is for device B, (c) is for device C, and (d) is for device D. The Figure 7e shows the output voltage at different excitation frequencies for the device D which represents dielectric-to-dielectric CS TENG.

Device A exhibited a rather chaotic output voltage with a change in excitation frequency, with almost no difference between the 6 Hz and 8 Hz excitation frequencies (see Figure 7a). For the excitation frequency of 10 Hz, the output voltage jumped to approximately 6 V peak-to-peak. Device B showed a smooth increase in output voltage with increasing excitation frequency. For an excitation frequency of 10 Hz, the output voltage was approximately 8.7 V peak-to-peak (Figure 7b). Device C also showed a smooth increase in output voltage with increasing excitation frequency. The magnitude of the output voltage was approximately comparable to device A; however, it was much steadier as the excitation frequency changed. Device C showed an output voltage of around 2.1 V peak-to-peak at an excitation frequency of 10 Hz (Figure 7c). Surprisingly, device D exhibited the highest steepness of output voltage increase at increasing excitation frequency. The output voltage was relatively stable over time for each excitation frequency. This device D exhibited a peak-to-peak output voltage of 11.5 V at an excitation frequency of 10 Hz (Figure 7d).

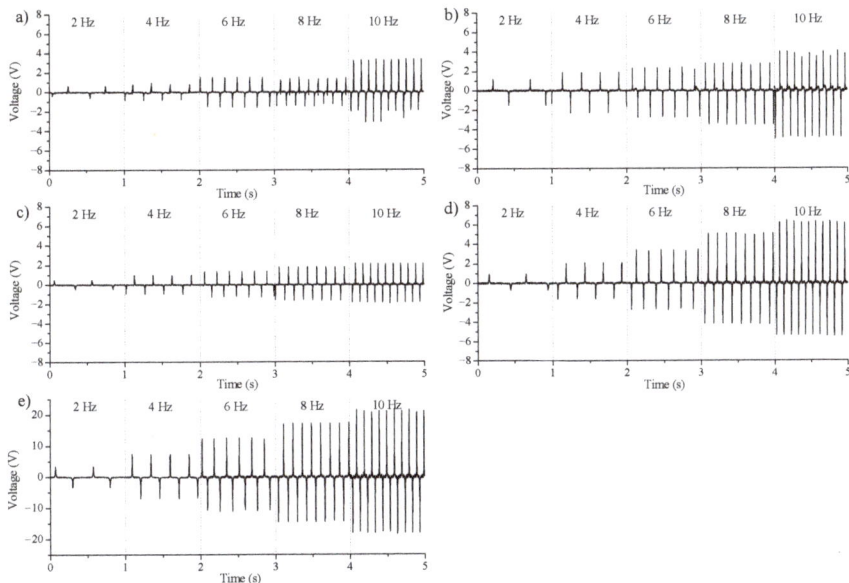

Figure 7. The output voltages for CS TENG at different excitation frequencies 2 Hz up to 10 Hz. The maximal gap between electrodes and force of touch of electrodes was constant. Individual TENGs are: (**a**) device A, (**b**) device B, (**c**) device C, (**d**) device D, and (**e**) device E.

In Figure 7e, it can be seen for the reference, how the output voltage behaved with increasing excitation frequency for the dielectric-to-dielectric CS TENG. It is clear to see that device E evidently generated the highest output voltage of the given TENG devices, where a peak voltage of 39 V was obtained at an excitation frequency of 10 Hz. However, device E represents an ideally assembled TENG when two materials with different triboelectric affinities are used.

It is necessary to note that this study was not performed for reaching the maximal power output of TENGs. However, the surface charge density (SCD) was also calculated for complete information about our TENGs. From the voltage dependencies measured at a known electrical resistance, the current delivered by each TENG can be derived. The current versus time for all TENGs is shown in Figure 8a. This dependence was measured for the excitation frequency of 10 Hz. By integrating the electric current over time, we have obtained the amount of electric charge delivered by each device. By averaging all the maximum and minimum charge values, the average value of the charge generated by the TENG during contact and separation was calculated. As the TENG has a size of (3×3) cm^2, the estimated surface charge density in the $\mu C/m^2$ unit is able to calculate by dividing the averaged charge by the device size. The surface charge density of all TENGs is shown in Figure 8b. Between the conductor-to-dielectric TENGs, device D generated the highest current and thus achieved the highest SCD ($I_{peak-to-peak} = 1.2\,\mu A$, $Q_{average} = 0.372$ nC). Followed by device B ($I_{peak-to-peak} = 0.9\,\mu A$, $Q_{average} = 0.279$ nC), where the SCD value was approximately 1.3× lower than that of device D. Even slightly lower values were observed for device A ($I_{peak-to-peak} = 0.6\,\mu A$, $Q_{average} = 0.202$ nC), and the lowest value was observed for device C ($I_{peak-to-peak} = 0.4\,\mu A$, $Q_{average} = 0.128$ nC), where the SCD was 2.9× lower than device D. It can also be seen from Figure 8, that for a given excitation, device E (representing the dielectric-to-dielectric TENG) generated the highest current, and thus achieved the highest SCD ($I_{\text{peak-to-peak}} = 3.9\,\mu A$, $Q_{\text{average}} = 1.235\,nC$) of all the TENGs. The generated charge was more than three times higher than that of device D, which exhibited the highest SCD value among the conductor-to-dielectric TENGs.

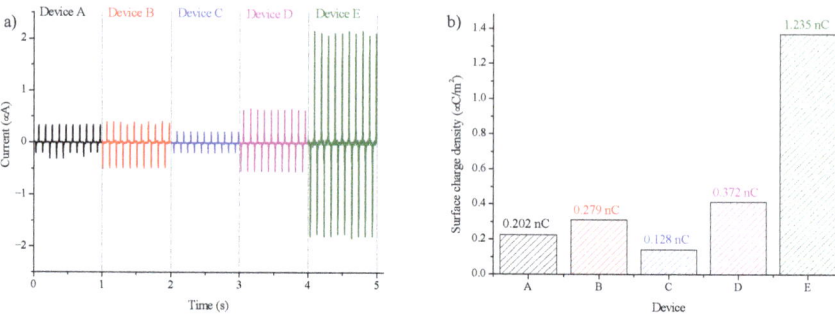

Figure 8. (**a**) The output current vs time for TENGs at excitation frequency 10 Hz. (**b**) The surface charge density of TENGs.

We can see that the SCD values achieved are very low. However, as mentioned above, this study was not based on achieving the maximum power of the TENGs. It was a study of two frequently used materials in TENGs, where one has a strong triboelectric affinity for negative charge (PVDF) and the other for positive charge (PA). These materials were prepared both separately in fibrous form and together in a fibrous composite, where the fiber thicknesses and the thickness of the overall layer were maintained.

A summary of the experimental data and their mutual comparison based on the device type is given in Table 2. It shows the type of triboelectric device, the structure used in the triboelectric device, the dielectric constant, and the dielectric loss at 1 kHz. The last column of Table 2 demonstrates a simplified comparison of the TENG sensitivities to the excitation signal.

Table 2. Summary of the experimental data.

Device	Type of the Triboelectric Device	Fiber Dielectric Sample	ε_r at 1 kHz	$\tan\delta\,(\times 10^{-3})$ at 1 kHz	Peak Voltage (V) at $F = 10\,N$, $f = 6\,Hz$	Triboelectric Response
A		PA	1.19	7.44	1.5	Lowest
B	Conductor-to-dielectric	PVDF	1.57	6.68	2.8	Medium
C		PVDF+PA	1.24	7.17	1.6	Low
D		S(PVDF+PA)	1.45	10.56	3.5	High
E	Dielectric-to-dielectric	PVDF opposite PA	1.57 PVDF 1.19 PA	6.68 PVDF 7.44 PA	14.2	Highest

Note: Thickness of the dielectric layer was 20 µm.

In addition, TENG devices were tested in terms of contact compression force, where the compression force was gradually increased and the output voltage was recorded from the TENGs. The excitation frequency of the compression was the same for all TENGs at 6 Hz. The measurement result for our TENGs is shown in Figure 9. The increasing output voltage with increasing contact compression force for all TENGs was observed. In this measurement, the device with the lowest sensitivity was device A. Such a device generated the lowest output voltage of all the TENGs. Device C exhibited slightly higher output voltage values than device A, where the peak output voltage was always about 1.4× higher than that of device A. This was followed by a relatively large jump, where device B exhibited about 2.5× higher peak output voltage compared to device A. The highest sensitivity of all conductor-to-dielectric CS TENGs was observed for device D, where the output voltage reached approximately 3.4 V at peak at a compression force (10 N). Again, for comparison, device E is also shown here, where was observed the highest sensitivity to contact force of all the TENGs used. This is due to the full contact and separation of the two fibrous materials PVDF and PA occurring. The triboelectric charge density σ of device E is higher than that of devices A, B, C and D. According to Equations (1), (3) and (4), the higher triboelectric sensitivity of device E (dielectric-to-dielectric TENG) can be predicted compared to devices A, B, C and D (conductor-to-dielectric TENG).

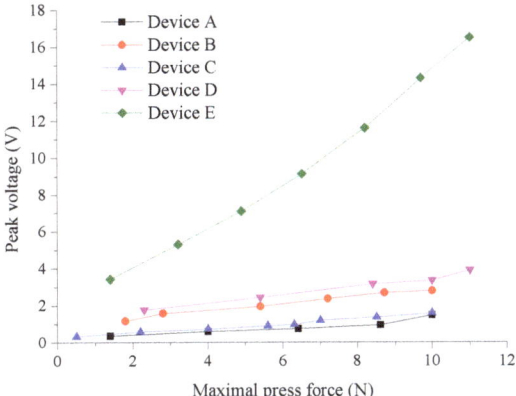

Figure 9. Peak voltage as a function of maximal press force for conductor-to-dielectric TENGs (device A, device B, device C and device D), and for dielectric-to-dielectric TENG, the device E.

A similar output voltage response was observed for device A (consisting of PA fibers) and device C (consisting of PVDF+PA intertwined fibers), which indicates that our PVDF fibers added no or only insignificant piezoelectric response. In the case of piezoelectrically active PVDF fibers, this device C would have to exhibit a higher voltage value than that observed in device A. Since this did not happen, it was concluded that mainly the triboelectric effect is applied here and the piezoelectric effect is suppressed.

This was confirmed by measuring the piezoelectric response on our PVDF fibers using the Berlincourt method, where PVDF fibers were sandwiched between two metal electrodes and measured by the d_{33} meter. This measurement confirmed zero or negligible piezoelectric response (maximum 1 pC/N) on our PVDF fibers.

The findings of this investigation yielded interesting results concerning the structure of the fibrous composite of the S(PVDF+PA) sample. Device D, formed by this composite, exhibited the highest triboelectric response among all conductor-to-dielectric TENGs. Although device D contained two materials with different triboelectric affinities (PVDF and PA), this device generated the highest voltage in triboelectric measurements compared to device A, device B and device C. The findings are novel and provide an interesting insight into building an efficient TENG that consists of only one triboelectric layer. The most efficient TENG consists of two materials with different triboelectric affinities that are completely separated from each other before each contact, as seen in device E. Device E is shown here for comparison purposes only and serves as an ideal case of the most efficient TENG assembly. However, it is sometimes impossible to form a TENG with two triboelectric layers. Then, this study shows that by combining both PVDF and PA materials into a single triboelectric layer, a TENG can be constructed that generates a higher voltage than that TENG formed by a single layer from PVDF or PA material. Another interesting finding is that, simply combining the two materials by intertwining them into a single triboelectric layer is not effective, as seen in the results measured on device C. The resulting triboelectric layer must also contain a trapping layer, which is formed by the PVDF fibers in our device D. By further optimizing the thickness of this sink layer, it would be possible to achieve an even higher efficiency of this single-layer TENG. However, this study is beyond the scope of this paper.

4. Conclusions

The novelty of this work lies in the study of the triboelectric properties of a fibrous composite material that contains two materials which lie at different ends in a triboelectric series. This composite, contains interwoven fibers of both materials then forms a single triboelectric layer in a triboelectric device. It should also be pointed out that this study was not focused on the maximum power output of the TENG. The study was focused on the possibility of using these composite fibrous materials as an active sensor. Comparing their triboelectric properties and whether the composite structure has any benefits. The experiment demonstrated that, to achieve the highest possible triboelectric sensitivity and the possibility of using a triboelectric device in energy harvesting applications, the individual triboelectric materials must be separated entirely during release before their next contact. This was shown on device E, where PVDF fibers were used against PA fibers, where we reached peak voltage of the device 14.2 V at excitation force 10 N and frequency 6 Hz. If the materials were still in partial contact during release and subsequent compression, the device operates in a sliding mode rather than a contact separation mode. The sliding mode can be used here for sensing applications better than for energy harvesting. This was demonstrated for device C, where PVDF fibers were intertwined with PA fibers. This triboelectric device C showed slightly higher sensitivity to the excitation signal than device A, which consisted of PA fibers. Device C generated the peak voltage of 1.6 V compared to device A, where the generated peak voltage was 1.5 V. Device A exhibited the lowest excitation signal sensitivity of all the assembled devices. Device D, which used a layered fibrous structure, showed the highest excitation signal sensitivity compared to the prepared conductor-to-dielectric triboelectric devices, where the measured voltage response was 3.5 V at peak. Such an improvement of the triboelectric response may be due to forming the trapping layer consisting of PVDF fibers placed between the electrode and the combined PVDF and PA layer. This device D exhibited 3.5× higher sensitivity to the excitation signal than device B composed of PVDF fibers, where the peak voltage 2.8 V was measured. The proposed study provides an interesting perspective on the fabrication of TENG formed by a single dielectric, where this dielectric contains a combination of two materials with different triboelectric affinities.

Author Contributions: Conceptualization, P.T. and J.K.; methodology, Z.H.; software, D.Ř.; validation, P.T., Z.H. and K.Č.; formal analysis, N.P.; investigation, D.S.; resources, P.T.; data curation, J.K.; writing—original draft preparation, P.T.; writing—review and editing, P.T.; visualization, N.P.; supervision, Z.H.; project administration, Ş.Ţ. All authors have read and agreed to the published version of the manuscript.

Funding: The research described in this paper was financially supported by the Grant Agency of Czech Republic under project No. 19-17457S. Part of the work was carried out with the support of CEITEC Nano Research Infrastructure supported by MEYS CR (LM2018110).

Institutional Review Board Statement: Not applicable.

Informed Consent Statement: Not applicable.

Conflicts of Interest: The authors declare no conflict of interest. The funders had no role in the design of the study; in the collection, analyses, or interpretation of data; in the writing of the manuscript; or in the decision to publish the results.

Sample Availability: Samples are available on demand from Pavel Tofel. E-mail: tofel@vut.cz.

Abbreviations

The following abbreviations are used in this manuscript:

AA	Acetic acid
Ac	Acetone
CS	Contact-separation
DMSO	Dimethylsulphoxide
FA	Formic acid
FT	Freestanding triboelectric-layer
LS	Lateral sliding
PA	Polyamide 6
PDMS	Polydimethylsiloxane
PVDF	Polyvinylidene fluoride
SE	Single-electrode
SEM	Scanning electron microscopy
SCD	Surface charge density

References

1. Xiong, J.; Lee, P.S. Progress on wearable triboelectric nanogenerators in shapes of fiber, yarn, and textile. *Sci. Technol. Adv. Mater.* **2019**, *20*, 837–857. [CrossRef]
2. Xia, K.; Wu, D.; Fu, J.; Hoque, N.A.; Ye, Y.; Xu, Z. A high-output triboelectric nanogenerator based on nickel–copper bimetallic hydroxide nanowrinkles for self-powered wearable electronics. *J. Mater. Chem. A* **2020**, *8*, 25995–26003. [CrossRef]
3. Zhang, P.; Zhang, Z.; Cai, J. A foot pressure sensor based on triboelectric nanogenerator for human motion monitoring. *Microsyst. Technol.* **2021**, *27*, 3507–3512. [CrossRef]
4. Heo, D.; Chung, J.; Kim, B.; Yong, H.; Shin, G.; Cho, J.W.; Kim, D.; Lee, S. Triboelectric speed bump as a self-powered automobile warning and velocity sensor. *Nano Energy* **2020**, *72*, 104719. [CrossRef]
5. Chen, T.; Shi, Q.; Li, K.; Yang, Z.; Liu, H.; Sun, L.; Dziuban, J.A.; Lee, C. Investigation of position sensing and energy harvesting of a flexible triboelectric touch pad. *Nanomaterials* **2018**, *8*, 613. [CrossRef] [PubMed]
6. Ibrahim, A.; Ramini, A.; Towfighian, S. Triboelectric energy harvester with large bandwidth under harmonic and random excitations. *Energy Rep.* **2020**, *6*, 2490–2502. [CrossRef]
7. Ghosh, R.; Pin, K.Y.; Reddy, V.S.; Jayathilaka, W.A.; Ji, D.; Serrano-García, W.; Bhargava, S.K.; Ramakrishna, S.; Chinnappan, A. Micro/nanofiber-based noninvasive devices for health monitoring diagnosis and rehabilitation. *Appl. Phys. Rev.* **2020**, *7*, 041309. [CrossRef]
8. Elahi, H.; Munir, K.; Eugeni, M.; Atek, S.; Gaudenzi, P. Energy harvesting towards self-powered iot devices. *Energies* **2020**, *13*, 5528. [CrossRef]
9. Chen, H.; Xing, C.; Li, Y.; Wang, J.; Xu, Y. Triboelectric nanogenerators for a macro-scale blue energy harvesting and self-powered marine environmental monitoring system. *Sustain. Energy Fuels* **2020**, *4*, 1063–1077. [CrossRef]
10. Gu, G.Q.; Han, C.B.; Lu, C.X.; He, C.; Jiang, T.; Gao, Z.L.; Li, C.J.; Wang, Z.L. Triboelectric Nanogenerator Enhanced Nanofiber Air Filters for Efficient Particulate Matter Removal. *ACS Nano* **2017**, *11*, 6211–6217. [CrossRef]
11. Ghatak, B.; Banerjee, S.; Ali, S.B.; Bandyopadhyay, R.; Das, N.; Mandal, D.; Tudu, B. Design of a self-powered triboelectric face mask. *Nano Energy* **2021**, *79*, 105387. [CrossRef] [PubMed]
12. Shi, L.; Jin, H.; Dong, S.; Huang, S.; Kuang, H.; Xu, H.; Chen, J.; Xuan, W.; Zhang, S.; Li, S.; et al. High-performance triboelectric nanogenerator based on electrospun PVDF-graphene nanosheet composite nanofibers for energy harvesting. *Nano Energy* **2021**, *80*, 105599. [CrossRef]
13. Rodrigues, C.; Nunes, D.; Clemente, D.; Mathias, N.; Correia, J.M.; Rosa-Santos, P.; Taveira-Pinto, F.; Morais, T.; Pereira, A.; Ventura, J. Emerging triboelectric nanogenerators for ocean wave energy harvesting: State of the art and future perspectives. *Energy Environ. Sci.* **2020**, *13*, 2657–2683. [CrossRef]
14. Zou, H.; Zhang, Y.; Guo, L.; Wang, P.; He, X.; Dai, G.; Zheng, H.; Chen, C.; Wang, A.C.; Xu, C.; et al. Quantifying the triboelectric series. *Nat. Commun.* **2019**, *10*, 1427. [CrossRef]
15. Wang, Z.L. Triboelectric nanogenerators as new energy technology for self-powered systems and as active mechanical and chemical sensors. *ACS Nano* **2013**, *7*, 9533–9557. [CrossRef]
16. Kim, Y.J.; Lee, J.; Park, S.; Park, C.; Park, C.; Choi, H.J. Effect of the relative permittivity of oxides on the performance of triboelectric nanogenerators. *RSC Adv.* **2017**, *7*, 49368–49373. [CrossRef]
17. Dhakar, L. *Triboelectric Devices for Power Generation and Self-Powered Sensing Applications*; Springer: Singapore, 2017. [CrossRef]
18. Částková, K.; Kaštyl, J.; Sobola, D.; Petruš, J.; Šťastná, E.; Říha, D.; Tofel, P. Structure–Properties Relationship of Electrospun PVDF Fibers. *Nanomaterials* **2020**, *10*, 1221. [CrossRef]
19. Sobola, D.; Kaspar, P.; Částková, K.; Dallaev, R.; Papež, N.; Sedlák, P.; Trčka, T.; Orudzhev, F.; Kaštyl, J.; Weiser, A.; et al. PVDF Fibers Modification by Nitrate Salts Doping. *Polymers* **2021**, *13*, 2439. [CrossRef]

20. Černohorský, P.; Pisarenko, T.; Papež, N.; Sobola, D.; Ţălu, Ş.; Částková, K.; Kaštyl, J.; Macků, R.; Škarvada, P.; Sedlák, P. Structure Tuning and Electrical Properties of Mixed PVDF and Nylon Nanofibers. *Materials* **2021**, *14*, 6096. [CrossRef]
21. Kaspar, P.; Sobola, D.; Částková, K.; Knápek, A.; Burda, D.; Orudzhev, F.; Dallaev, R.; Tofel, P.; Trčka, T.; Grmela, L.; et al. Characterization of Polyvinylidene Fluoride (PVDF) Electrospun Fibers Doped by Carbon Flakes. *Polymers* **2020**, *12*, 2766. [CrossRef]
22. Wang, W.; Wang, H.; Wang, H.; Jin, X.; Li, J.; Zhu, Z. Electrospinning preparation of a large surface area, hierarchically porous, and interconnected carbon nanofibrous network using polysulfone as a sacrificial polymer for high performance supercapacitors. *RSC Adv.* **2018**, *8*, 28480–28486. [CrossRef]
23. Chen, F.; Wu, Y.; Ding, Z.; Xia, X.; Li, S.; Zheng, H.; Diao, C.; Yue, G.; Zi, Y. A novel triboelectric nanogenerator based on electrospun polyvinylidene fluoride nanofibers for effective acoustic energy harvesting and self-powered multifunctional sensing. *Nano Energy* **2019**, *56*, 241–251. [CrossRef]
24. Garain, S.; Jana, S.; Sinha, T.K.; Mandal, D. Design of in Situ Poled Ce3+-Doped Electrospun PVDF/Graphene Composite Nanofibers for Fabrication of Nanopressure Sensor and Ultrasensitive Acoustic Nanogenerator. *ACS Appl. Mater. Interfaces* **2016**, *8*, 4532–4540. [CrossRef] [PubMed]
25. Mi, H.Y.; Jing, X.; Zheng, Q.; Fang, L.; Huang, H.X.; Turng, L.S.; Gong, S. High-performance flexible triboelectric nanogenerator based on porous aerogels and electrospun nanofibers for energy harvesting and sensitive self-powered sensing. *Nano Energy* **2018**, *48*, 327–336. [CrossRef]
26. Gasparini, C.; Aluigi, A.; Pace, G.; Molina-García, M.A.; Treossi, E.; Ruani, G.; Candini, A.; Melucci, M.; Bettin, C.; Bonaccorso, F.; et al. Enhancing triboelectric performances of electrospun poly(vinylidene fluoride) with graphene oxide sheets. *Graphene Technol.* **2020**, *5*, 49–57. [CrossRef]
27. Ţălu, Ş. *Micro and Nanoscale Characterization of Three Dimensional Surfaces: Basics and Applications*; Napoca Star Publishing House: Cluj-Napoca, Romania, 2015.
28. Zhang, F.; Li, B.; Zheng, J.; Xu, C. Facile Fabrication of Micro-Nano Structured Triboelectric Nanogenerator with High Electric Output. *Nanoscale Res. Lett.* **2015**, *10*, 4–9. [CrossRef]
29. Yu, J.; Hou, X.; He, J.; Cui, M.; Wang, C.; Geng, W.; Mu, J.; Han, B.; Chou, X. Ultra-flexible and high-sensitive triboelectric nanogenerator as electronic skin for self-powered human physiological signal monitoring. *Nano Energy* **2020**, *69*, 104437. [CrossRef]
30. Feng, Y.; Zheng, Y.; Zhang, G.; Wang, D.; Zhou, F.; Liu, W. A new protocol toward high output TENG with polyimide as charge storage layer. *Nano Energy* **2017**, *38*, 467–476. [CrossRef]
31. Kim, D.W.; Lee, J.H.; You, I.; Kim, J.K.; Jeong, U. Adding a stretchable deep-trap interlayer for high-performance stretchable triboelectric nanogenerators. *Nano Energy* **2018**, *50*, 192–200. [CrossRef]
32. Chen, A.; Zhang, C.; Zhu, G.; Wang, Z.L. Polymer Materials for High-Performance Triboelectric Nanogenerators. *Adv. Sci.* **2020**, *7*, 1–25. [CrossRef] [PubMed]
33. Garcia, C.; Trendafilova, I.; de Villoria, R.G.; del Rio, J.S. Self-powered pressure sensor based on the triboelectric effect and its analysis using dynamic mechanical analysis. *Nano Energy* **2018**, *50*, 401–409. [CrossRef]
34. Wu, C.; Wang, A.C.; Ding, W.; Guo, H.; Wang, Z.L. Triboelectric Nanogenerator: A Foundation of the Energy for the New Era. *Adv. Energy Mater.* **2019**, *9*, 1802906. [CrossRef]
35. Taghavi, M.; Beccai, L. A contact-key triboelectric nanogenerator: Theoretical and experimental study on motion speed influence. *Nano Energy* **2015**, *18*, 283–292. [CrossRef]
36. Niu, S.; Wang, S.; Lin, L.; Liu, Y.; Zhou, Y.S.; Hu, Y.; Wang, Z.L. Theoretical study of contact-mode triboelectric nanogenerators as an effective power source. *Energy Environ. Sci.* **2013**, *6*, 3576. [CrossRef]
37. Zhang, H.; Yao, L.; Quan, L.; Zheng, X. Theories for triboelectric nanogenerators: A comprehensive review. *Nanotechnol. Rev.* **2020**, *9*, 610–625. [CrossRef]
38. Smejkalová, T.; Ţălu, Ş.; Dallaev, R.; Částková, K.; Sobola, D.; Nazarov, A. SEM imaging and XPS characterization of doped PVDF fibers. *E3S Web Conf.* **2021**, *270*, 01011. [CrossRef]
39. Misiurev, D.; Ţălu, Ş.; Dallaev, R.; Sobola, D.; Goncharova, M. Preparation of PVDF-CNT composite. In *E3S Web of Conferences*; EDP Sciences: Les Ulis, France, 2021; Volume 270, p. 01012. [CrossRef]
40. Roscow, J.I.; Lewis, R.W.; Taylor, J.; Bowen, C.R. Modelling and fabrication of porous sandwich layer barium titanate with improved piezoelectric energy harvesting figures of merit. *Acta Mater.* **2017**, *128*, 207–217. [CrossRef]
41. Tong, J.; Zhang, H.; Li, W.; Chen, H.; Wang, D.; Hu, M.; Wang, Z. Simultaneously improving thermal conductivity and dielectric properties of poly(vinylidene fluoride)/expanded graphite via melt blending with polyamide 6. *J. Appl. Polym. Sci.* **2021**, *138*, 51354. [CrossRef]
42. Song, Y.; Shen, Y.; Hu, P.; Lin, Y.; Li, M.; Nan, C.W. Significant enhancement in energy density of polymer composites induced by dopamine-modified $Ba_{0.6}Sr_{0.4}TiO_3$ nanofibers. *Appl. Phys. Lett.* **2012**, *101*, 152904. [CrossRef]

Article

Early Recognition of the PCL/Fibrous Carbon Nanocomposites Interaction with Osteoblast-like Cells by Raman Spectroscopy

Aleksandra Wesełucha-Birczyńska [1,*], Anna Kołodziej [1], Małgorzata Świętek [2], Łukasz Skalniak [1], Elżbieta Długoń [3], Maria Pajda [4] and Marta Błażewicz [3]

1 Faculty of Chemistry, Jagiellonian University, Gronostajowa 2, 30-387 Kraków, Poland; anka.kolodziej@doctoral.uj.edu.pl (A.K.); lukasz.skalniak@uj.edu.pl (Ł.S.)
2 Institute of Macromolecular Chemistry, Czech Academy of Sciences, Heyrovského Sq. 2, 162 06 Prague, Czech Republic; swietek@imc.cas.cz
3 Faculty of Materials Science and Ceramics, AGH-University of Science and Technology, Mickiewicza 30, 30-059 Kraków, Poland; dlugon@agh.edu.pl (E.D.); mblazew@agh.edu.pl (M.B.)
4 Technolutions, Wiejska 7, 99-400 Łowicz, Poland; marynia85@gmail.com
* Correspondence: birczyns@chemia.uj.edu.pl; Tel.: +48-12-686-2772

Citation: Wesełucha-Birczyńska, A.; Kołodziej, A.; Świętek, M.; Skalniak, Ł.; Długoń, E.; Pajda, M.; Błażewicz, M. Early Recognition of the PCL/Fibrous Carbon Nanocomposites Interaction with Osteoblast-like Cells by Raman Spectroscopy. *Nanomaterials* **2021**, *11*, 2890. https://doi.org/10.3390/nano11112890

Academic Editor: Teresa Cuberes

Received: 7 August 2021
Accepted: 25 October 2021
Published: 28 October 2021

Publisher's Note: MDPI stays neutral with regard to jurisdictional claims in published maps and institutional affiliations.

Copyright: © 2021 by the authors. Licensee MDPI, Basel, Switzerland. This article is an open access article distributed under the terms and conditions of the Creative Commons Attribution (CC BY) license (https://creativecommons.org/licenses/by/4.0/).

Abstract: Poly(ε-caprolactone) (PCL) is a biocompatible resorbable material, but its use is limited due to the fact that it is characterized by the lack of cell adhesion to its surface. Various chemical and physical methods are described in the literature, as well as modifications with various nanoparticles aimed at giving it such surface properties that would positively affect cell adhesion. Nanomaterials, in the form of membranes, were obtained by the introduction of multi-walled carbon nanotubes (MWCNTs and functionalized nanotubes, MWCNTs-f) as well as electro-spun carbon nanofibers (ESCNFs, and functionalized nanofibers, ESCNFs-f) into a PCL matrix. Their properties were compared with that of reference, unmodified PCL membrane. Human osteoblast-like cell line, U-2 OS (expressing green fluorescent protein, GFP) was seeded on the evaluated nanomaterial membranes at relatively low confluency and cultured in the standard cell culture conditions. The attachment and the growth of the cell populations on the polymer and nanocomposite samples were monitored throughout the first week of culture with fluorescence microscopy. Simultaneously, Raman microspectroscopy was also used to track the dependence of U-2 OS cell development on the type of nanomaterial, and it has proven to be the best method for the early detection of nanomaterial/cell interactions. The differentiation of interactions depending on the type of nanoadditive is indicated by the ν(COC) vibration range, which indicates the interaction with PCL membranes with carbon nanotubes, while it is irrelevant for PCL with carbon nanofibers, for which no changes are observed. The vibration range ω(CH$_2$) indicates the interaction for PCL with carbon nanofibers with seeded cells. The crystallinity of the area ν(C=O) increases for PCL/MWCNTs and for PCL/MWCNTs-f, while it decreases for PCL/ESCNFs and for PCL/ESCNFs-f with seeded cells. The crystallinity of the membranes, which is determined by Raman microspectroscopy, allows for the assessment of polymer structure changes and their degradability caused by the secretion of cell products into the ECM and the differentiation of interactions depending on the carbon nanostructure. The obtained nanocomposite membranes are promising bioactive materials.

Keywords: nanomaterials; poly(ε-caprolactone) (PCL); multi-walled carbon nanotubes (MWCNTs); electro-spun carbon nanofibers (ESCNFs); Raman microspectroscopy; human U-2 OS cell line; bioactivity

1. Introduction

The most important task of regenerative medicine is to stimulate the body to carry out and accelerate the processes of self-repair of damaged cells and tissues [1,2]. The potential of regenerative medicine is related to the compilation of achievements in various fields, e.g., tissue engineering, genetics, biology, transplantology and materials engineering [1,3,4]. The task of material engineering is to design and manufacture substrates optimized to the needs and requirements of a given type of cells and tissue [5,6]. Therefore, materials planned for use in bone regenerative medicine should not only meet the conditions required for all biomaterials, i.e., biocompatibility, but also have the osteoinductive character and the ability to osseointegrate [7–9]. Another issue is to tailor the mechanical properties of the substrate to the natural bone parameters, and also biomimetically match at the macro, micro and nanoscopic level [10,11]. Natural substances such as collagen, cellulose, chitosan, alginic acid, bioceramics, biodegradable polymers and nanocomposites are readily used to prepare the bases of regenerative bone tissue [12–17].

PCL is a biocompatible resorbable material used in medicine, but its use is limited due to the fact that it is characterized by the lack of cell adhesion to its surface [10]. Various chemical and physical methods are described in the literature, as well as modifications with various nanoparticles aimed at giving it such surface properties that would positively affect cell adhesion [13]. The introduction of small amounts of nanoparticles in polymer matrices modifies the properties of polymers important for the applications in the field of regenerative medicine [18]. The type of nanoparticle introduced into the polymer matrix can cause not only a change in the polymer parameters, e.g., mechanical properties or thermal stability, but leads to a material with completely new properties, e.g., conductive or magnetic [19–22]. The adhesion of cells to the material surface depends on many factors, such as nano- and micro-scale topography, surface energy and certain mechanical properties—especially material stiffness, in particular—are the basic elements influencing the cellular response.

Carbon nanoforms (MWCNTs, CNFs, graphene) are materials with high potential for medical applications, not only in the area of their direct use (drug carriers, hyperthermia) but also in the field of surface or volume modifications of bioactive and biocompatible polymers [4,23–26]. Polymer nanocomposites containing carbon nanoparticles obtain a number of new functional properties that allow them to be used in nerve regeneration (nerve guide) or bone tissue (bioactive biomimetic scaffolds) applications [6,27,28]. The properties of polymer nanocomposites are closely related to the type of carbon nanoaddition and depend on the form, size and chemical structure of the surface [23,29,30]. Control of the properties of polymer nanocomposites due to their suitability for medical purposes requires a knowledge of phenomena at the molecular level accompanying the introduction of carbon nanoforms into the polymer matrix. We have carried out such tests for carbon layers on the titanium substrate [31,32], and we have also begun such studies for polymer nanocomposites [20]. It can be pointed out that the groups of atoms of the polymer matrix selectively interact with the nanoparticle of the nanoaddition.

In this work, nanomaterials in the form of membranes were produced by the introduction of multi-walled carbon nanotubes (MWCNTs) and functionalized multi-walled carbon nanotubes (MWCNT-f) as well as electro-spun carbon nanofibers (ESCNFs) and functionalized carbon nanofibers (ESCNFs-f) into a poly(ε-caprolactone) matrix (PCL). These nanomaterial membranes were brought into contact with the human osteoblast-like U-2 OS cell line and their interaction with the material was examined. The observed phenomena were compared with those observed for the reference polymer (PCL) membrane. The development of cell population, in the first days of culture, was monitored with fluorescence microscopy. Raman microspectroscopy was also applied to simultaneously verify interactions between the nanomaterials' phases, i.e., at the interface of the fibrous carbon-based nanoparticles and polymer, and also at the nanomaterial/cell interface. The interaction was analyzed in relation to the changes observed in the crystallinity of the polymer matrix and carbon nanoparticles as well as the U-2 OS cell response. This study enriches the information obtained so far by applying two-dimensional correlation [33,34]. The Raman microspectroscopy method is regarded as one of the new analytical approaches to study liquid/solid interfaces at the molecular level [35].

In this study, we present research on a modified polymer with MWCNTs and we compare these results with a material modified with a completely different carbon nanoform, which is ESCNFs, i.e., a material different from MWCNTs both in terms of crystalline structure and geometric parameters. In our approach to the analysis of nanocomposite membranes, we use the possibility of insight into interactions at the molecular level between a complex nanomaterial, i.e., certain molecular fragments components of a polymer matrix or carbon nanostructure, with osteoblast-like cells using Raman microscopy. In other words, we unravel the chemical changes that take place in cells in contact with four types of materials and correlate them with changes occurring within nanocomposites, as well as characterizing the phenomena occurring in carbon nanoforms. It is clear that recognizing the molecular properties of materials is important because they influence their macroscopic characteristics.

2. Materials and Methods

2.1. Fabrication of Nanocomposite Membranes

Poly(ε-caprolactone (PCL; $(C_6H_{10}O_2)_n$, Mn 45.000; purchased from Sigma-Aldrich, Warsaw, Poland), designed as a matrix, was dissolved in dichloromethane (DCM; provided by Avantor Performance Materials, Gliwice, Poland) to prepare its 10 wt% solution which was stirred overnight at room temperature. The nanoadditives, namely multi-wall carbon nanotubes (MWCNTs) or functionalized multi-wall carbon nanotubes (MWCNTs-f), electro-spun carbon nanofibers (ESCNFs) or functionalized carbon nanofibers (ESCNFs-f), were dispersed in an equal volume of organic solvent with an aid of the sonication process—firstly in an ultrasonic bath (L&R Manufacturing Co., Kearny, NJ, USA) for 10 min and then additionally by using a sonication probe for 3 min at an amplitude of 30% (BANDELIN electronic GmbH & Co. KG, Berlin, Germany).The obtained suspension was immediately transferred into the polymer solution. The mixture was sonicated for 3 min at an amplitude of 30% to ensure a good combination of both constituents, promptly poured onto a Petri dish (diameter 55 mm) and left at room temperature. The Petri dish was protected with punctured foil from too high rate of DCM evaporation. The produced polymer nanocomposites, PCL/MWCNTs and PCL/MWCNTs-f, PCL/ESCNFs or PCL/ESCNFs-f, contained 0.5 wt% nanoadditive in each material in relation to the weight of the polymer. The manufacturing process is shown in Figure 1, and was also described previously [33,34].

The MWCNTs (obtained from Nanostructured & Amorphous Materials, Inc., USA; purity: $\geq$95%; length: 0.5–2 µm; outside diameter: 10–30 nm) were functionalized in a mixture of sulphuric (VI) acid and 65% nitric (V) acid with 3/1 ratio 70 °C for 2 h (functionalized multi-walled carbon nanotubes, MWCNTs-f). Then, the nanotubes were rinsed with distilled water and centrifuged (Figure 1a) [20]. The total oxygen content in the MWCNTs is about 7%, which indicates a relatively small degree of functionalization of the tested nanotubes; for MWCNT-f it was estimated twice as much [36].

Carbon nanofibers were produced in the course of carbonization of a polyacrylonitrile (PAN) precursor, consisting of copolymers, 93–94 wt% acrylonitrile, 5–6 wt% methyl methacrylate and 1 wt% sodium allylsulfonate (purchased from Mavilon, Hungary). The PAN nanofibers were spun from a 11% solution N'N-dimethylformamide (DMF, acquired from Avantor Performance Materials Poland S.A.) using an electrospinning setup consisting of a high voltage generator (regulated from 1 to 20 kV), rotating tubular collector and a syringe with the polymer solution with a nozzle made of a stainless-steel needle with a diameter in the range of 0.6–1.2 mm. Prior to the process of electrospinning, the solution was stirred with a magnetic stirrer for 24 h. The average diameter of a nanofibers thus obtained was 250–280 nm. The as-obtained precursor was turned into carbon fibers in a three-step process. The first step was thermo-oxidative stabilization at a temperature of 250 °C and was performed for 1 h. The oxidation process was expected to transform the linear structure of the polymer into a cyclic structure. Then followed low-temperature (750 °C, 1 h) and high-temperature carbonization (1000 °C, 1 h) conducted in the protective atmosphere of nitrogen flow (30 L/h), with a heating rate of 5 °C per minute [37]. Carbon nanofibers were subjected to oxidation treatment in concentrated nitric acid (V) at 65 °C for 1 h (Figure 1b). Then samples were cooled down in the solution to room temperature, washed and dried in a dryer [38].

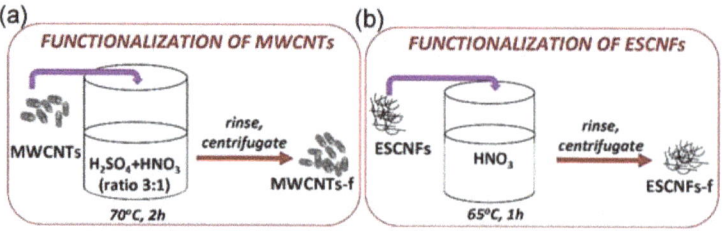

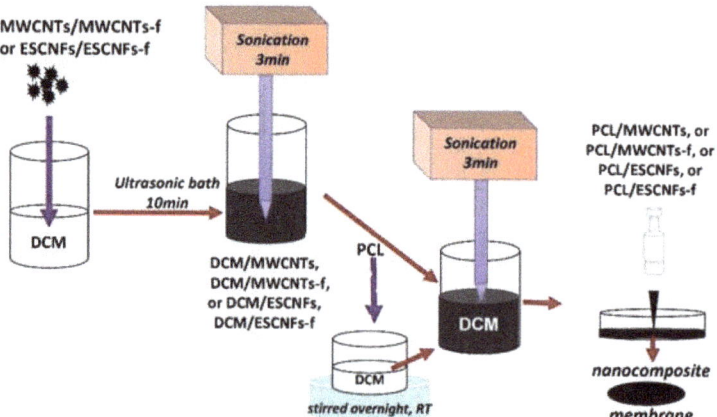

Figure 1. The nanocomposite membranes fabrication process; the inserts present a method of functionalization of carbon nanoadditives: (**a**) functionalization of MWCNTs; (**b**) functionalization of ESCNFs.

The obtained membranes are shown in Figure 2a,c–f.

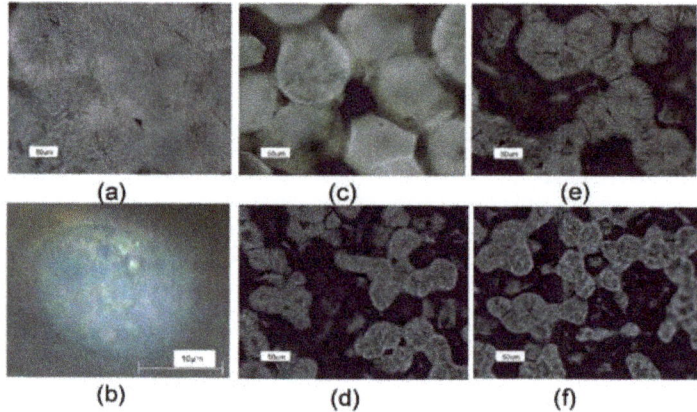

Figure 2. Microphotograph of the membrane top face: (**a**) PCL; (**c**) PCL/MWCNTs; (**d**) PCL/ESCNFs; (**e**) PCL/MWCNT-f and (**f**) PCL/ESCNFs-f, magnification 20×; (**b**) U-2 OS human cell on PCL/MWCNTs on the first day of culture, immersion objective, magnification 60×.

2.2. Contact Angle Measurements and Surface Free Energy Evaluation

The contact angle measurements were performed on a SAM10Mk1 (KRÜSS GmbH, Germany) goniometer using deionized water, by the sessile drop method. In order to determine the surface free energy (SFE) contact angle for non-polar diiodomethane (CH_2I_2) was measured, additionally to water (the polar liquid). A calculation model according to Owens, Wendt, Rabel and Kaelble (OWRK-model) available for SFE within the factory-supplied software was employed. The calculation requires the contact angles of two liquids with known polar and diffuse SFE fractions. Then, the free energy of a surface can be considered as composed of the polar part and the dispersed part. At least ten contact angle measurements in different locations on the surface were performed to obtain an average value. The results of the contact angle measurements and surface free energy for all membranes were statistically analyzed by calculating the arithmetic mean of the results and the standard population deviation function in Excel software. The Kolmogorov–Smirnov test of normality was performed, $p < 0.001$.

2.3. Cell Culture

The human U-2 OS cell line (ECACC, cat. no. 92022711, lot no. 10K035) is one of the first generated cell lines from the moderately differentiated osteosarcoma and is used quite frequently to test materials bioactivity [39]. The cells were cultured in Mc Coy's medium (BioWest, Nuaillé, France) supplemented with 10% Fetal Bovine Serum (FBS, BioWest, Nuaillé, France). The cells were grown at 37 °C in a humified atmosphere containing 5% CO_2.

The monoclonal population of cells with stable expression of maxFP-Green, a tailored green fluorescence protein, was developed from the U-2 OS cell line by transfection with the pmaxFP-Green-N vector (Amaxa Biosystems, Cologne, Germany). The transfection was completed using Lipofectamine 2000 (Life Technologies, Carlsbad, CA, USA), and stable clones were selected with G418 (Life Technologies). The resulting clones were picked, repopulated and verified for the maxFP-Green expression with the flow cytometry. A clone, designated U-2 OS-Green, had optimal expression of a transgene and was used for the experiments.

The nanomaterial samples were sterilized in 70% ethanol for 30 min. After washing three times with saline phosphate buffer solution (PBS), they were exposed to UV light for 30 min. For the experiments, the cells were seeded at a relatively low density (10,000 cells per cm^2) in 12-well plates. The next day, the materials were transferred into new 12-well plates with a fresh cell culture medium, in order to exclude the cells growing on the plastic from imaging. The procedure was described previously [33,34].

2.4. Fluorescence Microscopy

The development of the fluorescent U-2 OS-Green on the analyzed nanomaterials was monitored using a Leica DM IL Led fluorescence microscope (Leica Microsystems, Wetzlar, Germany), equipped with a Leica DFC3000 G digital camera(Leica Microsystems, Wetzlar, Germany). The images were captured with Leica Application Siute X 3.3.3.16958 software using the Leica N PLAN 10x/0.25 PH1 objective, and analyzed with ImageJ 1.48v [40]. While the cells were seeded and grown on the upper surface of the materials, just before the imaging the materials were inverted, and inverted back after the imaging for further culture. The growth of the cells was monitored at the 1st, 2nd, 3rd and 6th day post-seeding and every day the images were captured using the same camera settings (40 ms exposure, gain = 1) to enable quantification of the fluorescence intensity (day 6 was an exception, when the 15 ms exposure was performed to capture the properly exposed image, but this was compensated in the calculations).

The information on the number of photomicrographs was analyzed and the statistical analysis is provided in the captions in Figures 3 and 4. For fluorescence quantification, background subtraction was performed for each individual image in order to provide more adequate data and better reflect the differences observed in the photomicrograph.

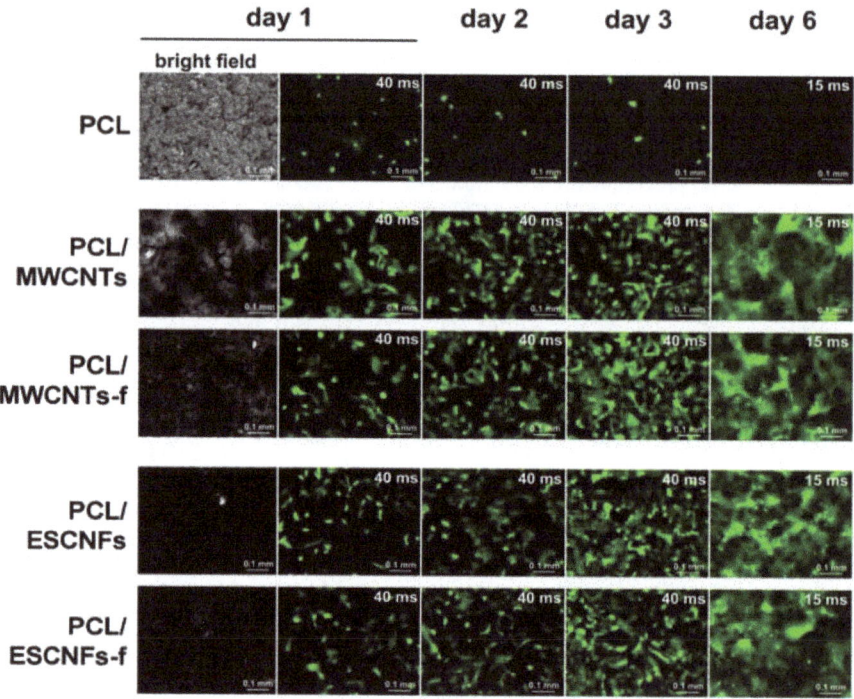

Figure 3. U-2 OS-Green cells expressing green fluorescent protein (GFP) in bright field and cultured for the 1st, 2nd, 3rd (40 ms exposure, gain = 1) and 6th day post-seeding (15 ms exposure, gain = 1), from top: PCL; PCL/MWCNTs, PCL/MWCNTs-f, PCL/ESCNFs and PCL/ESCNFs-f; mag. 10× (original images were published in [33,34]). The exposure time on the sixth day was reduced due to the appropriate capture of the exposed image. The cell growth experiments were performed at least three times for each material.

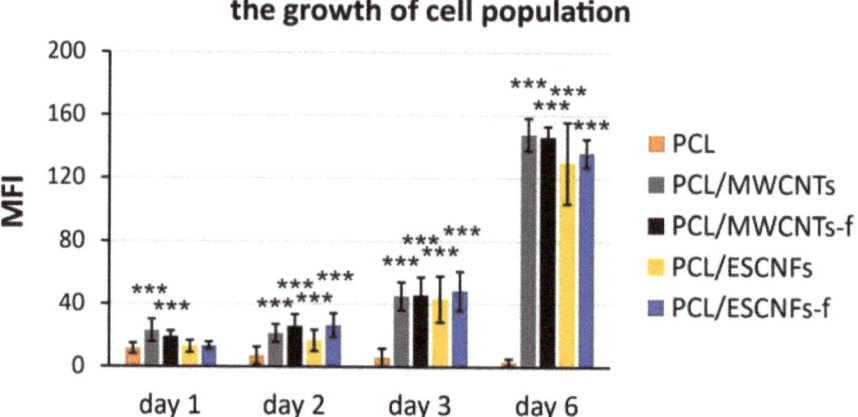

Figure 4. The growth of U-2 OS-Green cells on PCL; PCL/MWCNTs, PCL/MWCNTs-f, PCL/ESCNFs and PCL/ESCNFs-f; nanomaterials, monitored at the 1st, 2nd, 3rd (40 ms exposure, gain = 1) and 6th day post-seeding (15 ms exposure, gain = 1). The bar represents the means ± SD of the mean fluorescence intensities (MFI) from 9–20 separate images (median images number = 12.5). For the MFI quantifications background subtraction was applied. Statistical analysis was performed with one-way ANOVA with the Tukey post-hoc test: ***, $p < 0.001$ versus PCL. The differences in exposure time were compensated in the calculations.

2.5. Raman Microspectroscopy

A Renishaw inVia spectrometer (Wotton-under-Edge, Gloucestershire, UK), working in a confocal mode, connected to a Leica microscope (Leica Microsystems, Wetzlar, Germany), was used for the measurement of the Raman spectra. The beam from a 785 nm HP NIR (high power near IR) diode laser was focused on the samples by a Nikon immersive objective 60× magnifying (NA = 0.5). Raman light was dispersed by diffraction grating with 1200 grooves/mm. Laser power was kept low, c.a. 1–3 mW on the sample, to ensure minimum disturbances of the samples. The Raman spectra of the studied PCL/MWCNTs/ MWCNTs-f and PCL/ESCNFs/ ESCNFs-f nanomaterial membranes and reference PCL membrane cultured with U-2 OS cells were collected at the 1st, 3rd, 6th and 8th day post-seeding. Measurements of membranes with cells seeded on their surface were recorded in the range of 2000–400 cm^{-1} to shorten the measurement time to reduce cell signal disturbance. Four accumulations were made for each measurement site. The spectra were averaged by adding five spectra to thereby also improve the signal-to-noise ratio. Statistical analysis of Raman spectra was carried out in the OMNIC program, the average position and the standard deviation resulting from the summation.

Factory-supplied software was used to preprocess, i.e., cosmic spike removal, smooth and a baseline corrected (Renishaw, WiRE v. 2.0 and 3.2). The height, a width and percentage of the Lorentz–Gauss curve were fitted. From matching, the band parameters, the positions of the component bands, its height, full width at half-height (FWHH) and their area were determined. The curve fit procedure allowed for the analysis of changes in the marker areas characterizing regions of the matrix polymer chains, carbon nanoadditives and cells, through appropriate band intensity ratios. Changes could be determined by comparison with reference spectra and the corresponding reference intensity ratios for: the polymer matrix, nanocarbon additive and cell. Statistical analysis was performed with PCA with Calibration 99.30505%; Validation 97.97677% (the first measurement day); Calibration 97.08938%; Validation 95.18859% (the third measurement day); Calibration 99.35546%; Validation 98.39545% (the sixth measurement day); Calibration 97.8542%; Validation 94.61504% (the eighth measurement day).

3. Results and Discussion

3.1. The Morphology of Membranes of PCL with Fibrous Carbon Nanoparticles

The micrographs taken from the top face of the nanocomposite membranes and reference PCL membrane are shown in Figure 2. The main factor limiting the growth of a single spherulite is the growth of other spherulites in its immediate vicinity. The well-formed spherulites typical for PCL polymer of radius ~95 μm (Figure 2a) become smaller along with functionalization. Their radius is equal to ca. 70, 40, 25 and 23 μm for PCL/MWCNTs, PCL/MWCNTs-f, PCL/ESCNFs and PCL/ESCNFs-f, respectively. Additionally, significant changes in the surface morphology of the polymeric membranes are observed (Figure 2b–e). On the basis of the microphotographs it can be assumed that both unmodified fibrous carbon nanoforms and also, respectively, functionalized nanoparticles constitute the nucleation centers when introduced into the polymer matrix solution. The simultaneous crystallization of spherulites in many sites, combined with a limited possibility of their recrystallization, leads to the formation of numerous pores. As a consequence, the spherulitic structure of the material gradually disappears along with the increasing number of heterogeneous seeds of the crystallization.

3.2. Contact Angle Measurements and Surface Free Energy (SFE)

Wettability was determined at room temperature. Although the introduction of fibrous carbon nanoparticles into the polymer matrix resulted in a slight decrease in the nanomaterial membranes hydrophobicity, the calculated values of the contact angle for the tested materials are quite similar. Based on previous research. the values of the wetting angle for the top membranes surface are equal to 94.7 ± 1.2; 88.8 ± 1.3; 90.6 ± 3.7, 89.5 ± 1.5, 89.4 ± 1.2 for PCL, PCL/MWCNTs, PCL/MWCNTs-f, PCL/ESCNFs and

PCL/ESCNFs-f, respectively [41]. The interaction of the liquid phase with the materials occurs through polar and dispersion forces. SFE values are relatively large compared to other polymeric materials, but comparable for the tested membranes. Interestingly, the polar SFEs components for PCL/MWCNTs/MWCNTs-f are slightly higher than that for PCL/ESCNFs/ESCNFs-f, Table 1. Surface energy is the result of many factors; however, it cannot be ruled out that in this case the surface morphology will have a key impact on the parameters of the analyzed materials [4]. Perhaps the size of the spherulites affects the properties of the materials.

Table 1. Contact angle for diiodomethane and surface free energy for the top surfaces (the contact surface with the U-2 OS cells) of the studied membranes.

Material	Contact Angle for Diiodomethane [°]		Surface Free Energy [mN/m]		Disperse Part [mN/m]		Polar Part [mN/m]	
	Value	StDev	Value	StDev	Value	StDev	Value	StDev
PCL	29.31	2.76	45.83	1.42	44.51	1.12	1.32	0.29
PCL/MWCNTs	26.58	4.80	47.66	2.36	45.57	1.80	2.09	0.56
PCL/MWCNTs-f	25.98	3.07	46.70	1.60	45.83	1.13	0.88	0.48
PCL/ESCNFs	29.31	3.20	45.93	1.62	44.51	1.30	1.43	0.32
PCL/ESCNFs-f	35.71	3.02	42.88	2.02	41.70	1.42	1.19	0.60

3.3. The Comparison of Growth of U-2 OS Cells on the Membranes of PCL with Fibrous Carbon Nanoparticles

To test the ability of the materials to serve as a substrate of cell growth, human U-2 OS-Green cells were seeded on the tested nanomaterials, PCL/MWCNTs, PCL/MWCNTs-f and PCL/ESCNFs, PCL/ESCNFs-f, and also as a reference on the PCL membrane [33,34]. The cells were seeded at a relatively low confluency and cultured in standard cell culture conditions. This procedure is used to assess the properties of materials in contact with living cells [24–36]. It is common practice to use human osteogenic sarcoma cells (e.g., U-2 OS) cultured in vitro to investigate the biocompatibility of materials [37]. The growth of the cells was monitored by fluorescence microscopy at the 1st, 2nd, 3rd and 6th day post-seeding.

As we have described in our previous studies, for all of the studied nanocomposite substrates, a marked increase in the cell population was observed in the first week of culture while for a reference PCL membrane no such increase was observed [33,34]. In this manuscript, the cell growth was compared between all four nanocomposite membranes and the PCL in a single analysis (Figure 3). The growth was estimated by employing a quantitative method in fluorescence microscopy, i.e., quantifying mean fluorescence intensities for every captured image, which is proportional to the cell population numbers, calculated as the mean pixel intensity [42]. The analysis revealed a clear, exponential increase in the number of cells, starting from the second day of the culture, and observed on the consecutive days (Figure 4). This increase was estimated, by comparing the first and sixth day of culture, as equal to 3.5, 4.8, 3.9 and 4.0 for PCL/MWCNTs, PCL/MWCNTs-f, PCL/ESCNFs and PCL/ESCNFs-f, respectively. This suggests outstanding proliferation of the cells on the tested nanomaterials. In contrast, the numbers of cells seeded on a reference PCL membrane even slightly decreased with time, which may partially be a consequence of cell detachment during the material inverting procedure performed for microscopic visualization, and suggests that the modification of the PCL surface with nanoforms of carbon significantly improves the cells' attachment to the materials (Figure 4).

3.4. Raman Microspectroscopic Analysis of the Membranes of PCL with Fibrous Carbon Nanoparticles/Cells Interactions

Raman spectroscopy was used to monitor the interactions of polymer (PCL)-based carbon, fibrous nanomaterial membranes with human osteoblast-like U-2 OS cells at the 1st, 3rd, 6th and 8th day post-seeding. This interaction was monitored by analyzing changes in the crystallinity of the polymer matrix, by an identification of the ordering of the respective carbon nanoforms and recognition of the osteoblast U2-OS cells' marker bands. Figure 5 shows the Raman spectra on the first and last day of the experiment.

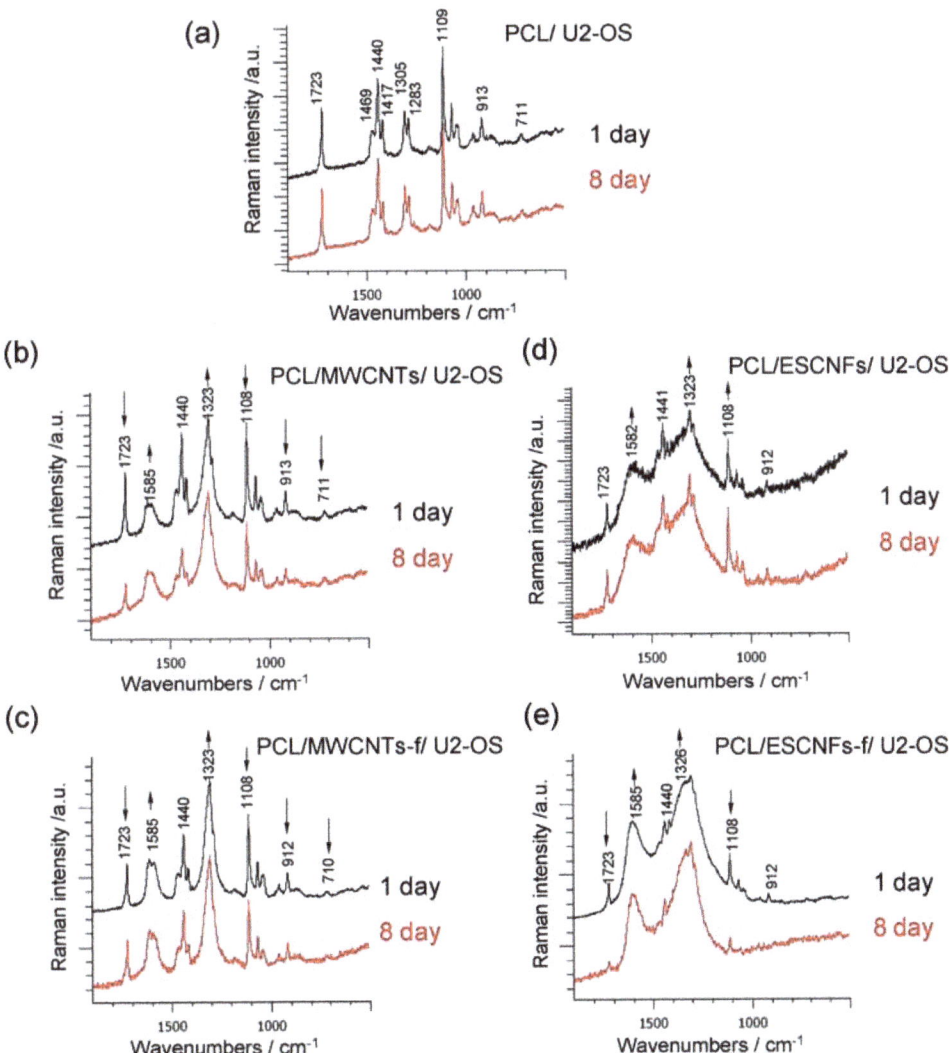

Figure 5. Raman spectra of the U-2 OS cell culture on the substrate: (**a**) PCL; (**b**) PCL/MWCNTs; (**c**) PCL/MWCNTs-f; (**d**) PCL/ESCNFs and (**e**) PCL/ESCNFs-f, on day 1st and 8th; 1900–500 cm^{-1} range, 785 nm excitation line.

3.4.1. PCL Matrix Crystallinity

The observed significant Raman bands and their assignments are collected in Table 2. The intensity of some marker bands characterizing the polymer crystallinity, i.e., stretching vibrations at 1723 cm^{-1} due to the ν(C=O), 1108 cm^{-1} band assigned to ν(COC), 913 cm^{-1} to ν(C-COO), and also the deformation vibrations at δ(CH$_2$) at 1440 and 1417 cm^{-1}, marked with arrows in Figure 5b–e, changed significantly in the first days of the culture. The evolution of the changes taking place in the tested nanomaterials in the consecutive measurement days was assessed in relation to selected markers of the polymer crystallinity, by matching lines and analytically determining the component bands in the appropriate ranges, and presented in Figure 6 [20,43].

Table 2. Observed characteristic Raman bands [cm^{-1}] and their assignments for PCL, PCL/MWCNTs, PCL/MWCNTs-f and PCL/ESCNFs and PCL/ESCNFs-f nanocomposite membranes in the first day of culture with human U-2 OS cell line, 785 nm laser line.

Raman Bands [cm^{-1}]					Assignment
PCL	PCL/ MWCNTs	PCL/ MWCNTs-f	PCL/ ESCNFs	PCL/ ESCNFs-f	
712 ± 1	712 ± 1	713 ± 2	713 ± 1	713 ± 2	$\delta(CH_2)$, $\delta(NH_2)$, Gly; CS, Cys [44–51]
865 ± 1	862 ± 1	862 ± 2	861 ± 1	861 ± 2	ν(C-COO) PCL (amorph); o.o.p. $\delta(CH2)$, Pro; collagen [20,43,45,52]
913 ± 1	913 ± 1	912 ± 1	912 ± 1	912 ± 1	ν(C-COO), PCL (cryst); $\tau(CH_2)$&$\tau(NH_2)$, Gly; collagen [20,43,44,52]
958 ± 1	958 ± 1	957 ± 1	958 ± 1	958 ± 1	ν(C-COO), PCL; ring str., Pro [20,43,45]
1038 ± 1	1038 ± 1	1037 ± 1	1038 ± 1	1037 ± 1	ν(COC), PCL; $\omega(CH_2)$, Pro; ν(CN)&ν(CC), Gly [20,43–45]
1064 ± 1	1064 ± 1	1064 ± 1	1064 ± 1	1064 ± 1	ν(COC), PCL (amorph) [20,43]
1109 ± 1	1109 ± 1	1108 ± 1	1109 ± 1	1108 ± 1	ν(COC), PCL (cryst); collagen [20,43,52]
1284 ± 1	1284 ± 1	1284 ± 1	1283 ± 1	1283 ± 1	$\omega(CH2)$, PCL (cyst); δ(CH2), Pro [20,43,45]
1305 ± 1	1305 ± 1	1305 ± 1	1305 ± 1	1304 ± 1	$\omega(CH2)$, PCL (cryst and amorph) [20,43]
-	1323 ± 1	1323 ± 1	1340 ± 2	1341 ± 2	D1-disorder-induced A$_{1g}$ mode in graphite plane; $\delta(CH_2)$, Pro [45,53,54]
1418 ± 1	1418 ± 1	1418 ± 1	1418 ± 1	1418 ± 2	δ(CH2), PCL; γ(CH2)), Gly [20,43,44,52]
1441 ± 1	1441 ± 1	1441 ± 1	1441 ± 1	1441 ± 1	δ(CH2), PCL (cryst.); δ(CH2), Pro [20,43,45]
1469 ± 1	1468 ± 1	1467 ± 1	1466 ± 1	1470 ± 1	δ(CH2)), PCL; collagen [20,43,52]
-	1587 ± 1	1585 ± 1	1584 ± 1	1585 ± 1	corresponding to G-graphite tangential mode [30,31,53]
-	1615 ± 1	1614 ± 1	1615 ± 1	1614 ± 1	D2-band due to due to in-plane defects and heteroatoms [54]
1723 ± 1	1723 ± 1	1723 ± 1	1724 ± 1	1723 ± 1	ν(C=O), PCL (cryst) [20,43]
1732 ± 1	1733 ± 1	1733 ± 1	1732 ± 1	1723 ± 1	ν(C=O), PCL (amorph) [20,43]

The intensity ratio of 1108 (cryst)/1097 (amorph) cm^{-1} ν(COC) vibrations in the PCL chain decreases in the first days of culture for all types of membranes, which indicates a decrease in the crystallinity of the polymer matrix, and then its increase on day 8 (Figure 6a,b). The increase in crystallinity on the 8th day of culture seems to indicate the stabilization of the polymer matrix in the process of cell adhesion related to their intense proliferation. It cannot be ruled out that the increase in proliferation may affect the growth of the band contribution of about 1100 cm^{-1}, for lipids and DNA, O-P-O backbone stretching, although this band for U-2 OS is not very intense in our measurement conditions [34,51,55]. The characteristics of the adjacent spectral regions give the intensity ratio of 913 (cryst)/864 (amorph) cm^{-1} due to ν(C-COO) vibrations, which for PCL/MWCNTs-f is similar to the previous ones, but for the PCL/MWCNTs it decreases in consecutive days (Figure 6c). However, for the PCL/ESCNF and PCL/ESCNF-f crystallinity of the polymer matrix it does not change significantly (Figure 6d). Variability in the C-C region appears to indicate that cell adhesion is taking place.

The band at 1305 cm^{-1} due to $\omega(CH_2)$ comes from the crystalline and amorphous PCL domains, while 1285 cm^{-1} originates only from $\omega(CH_2)$ in crystalline areas. The intensity ratio of 1285/1305 cm^{-1} decreases for PCL/MWCNTs, while it increases for PCL/ESCNF and PCL/ESCNF-f (Figure 6e,f). Interaction with cells seems to influence this process. The oscillation range $\omega(CH_2)$ indicates an increase in the interaction of PCL/ESCNFs and PCL/ESCNFs-f with cultured cells, observed by a systematic increase in the amorphousness in the studied system [15]. The influence of proliferating cells cannot be excluded, so that the Pro signal of approx. 1280 cm^{-1} increases the intensity ratio 1285/1305 [45].

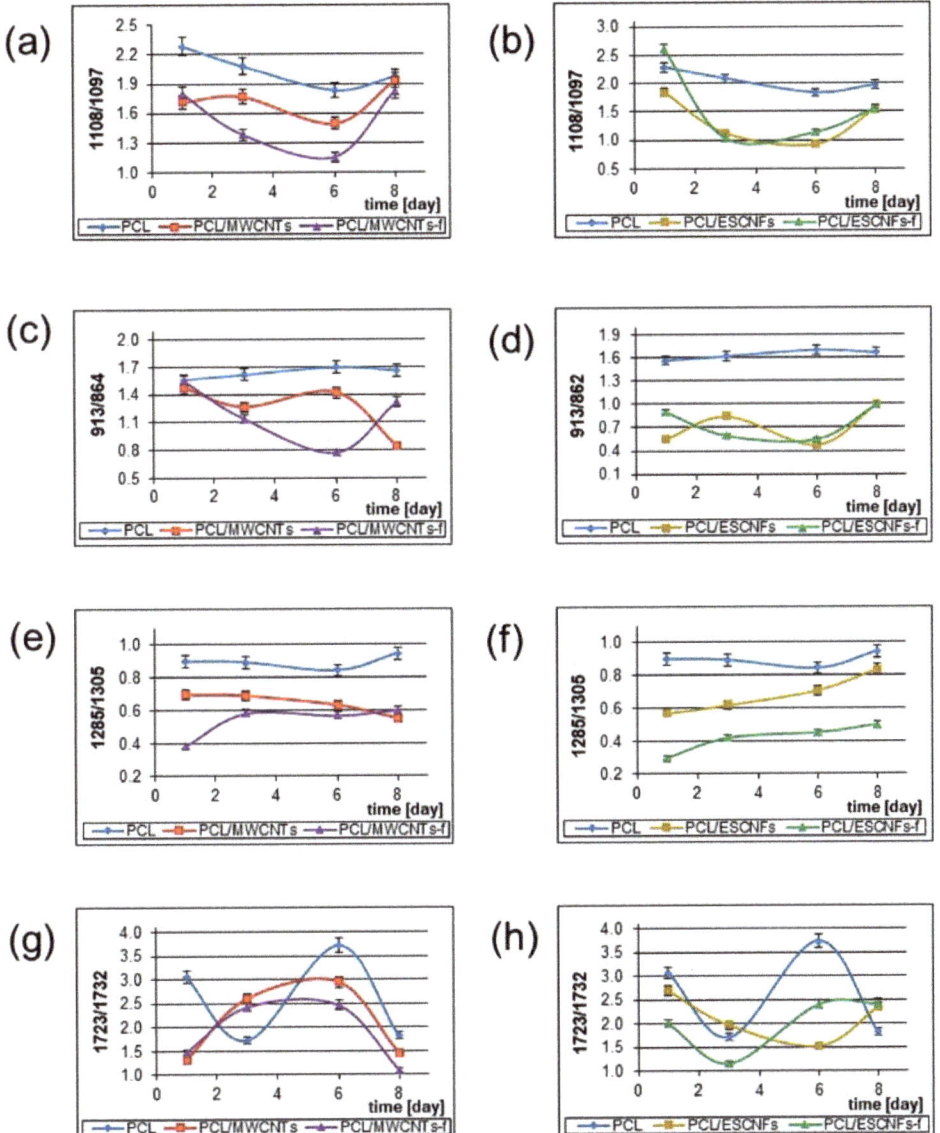

Figure 6. The intensity ratio (integral) of the PCL marker bands characterizing the crystallinity of the tested polymer nanomaterials constituting the substrate for U-2 OS cell culture, in the following days of culture. For PCL, PCL/MWCNTs, PCL/MWCNTs-f: (**a**) 1108/1097 cm^{-1} (Statistical analysis with PCA with Calibration 99.8774%; Validation 99.50629%); (**c**) 913/864 (Statistical analysis with PCA with Calibration 99.12821%; Validation 93.47666%); (**e**) 1285/1305 (Statistical analysis with PCA with Calibration 99.94765%; Validation 99.68526%); (**g**) 1723/1732 (Statistical analysis with PCA with Calibration 99.97593%; Validation 99.81823%) and for PCL, PCL/ESCNFs, and PCL/ESCNFs-f: (**b**) 1108/1097 (Statistical analysis with PCA with Calibration 99.8774%; Validation 99.6284%); (**d**) 913 /864 (Statistical analysis with PCA with Calibration 99.7179%; Validation 97.28348%); (**f**) 1285/1305 (Statistical analysis with PCA with Calibration 99.94656%; Validation 99.65956%); (**h**) 1723/1732 (Statistical analysis with PCA with Calibration 99.98973%; Validation 99.85501%), variance is defined by y-axis error bars, OMNIC software.

Another important parameter of crystallinity of the polymer matrix is the intensity ratio of 1723 (cryst)/1732 (amorph) cm^{-1} reflecting involvement of the C=O group in interactions with cells, which grows for PCL/MWCNTs and PCL/MWCNT-f, but at the 8th day significantly decreases (Figure 6g). It is different for PCL/ESCNF, for which the intensity ratio decreases and then increases, while for PCL/ESCNF-f the behavior is directly opposite (Figure 6h).

A decrease in the 1723/1732 intensity ratio, observed for PCL/MWCNTs membrane, indicates a decrease in the crystallinity of the polymer matrix in the area of C=O groups of the polymer chain, on the 8th day of culture. This indicates the influence of cells on their growth substrate, which occurs through the enlargement of the amorphous domains in the material. This evaluation is consistent with the results of a two-dimensional correlation analysis [34].

A similar pattern, the reduction of I1723/I1732 intensity ratio, was observed for the PCL/MWCNTs-f membrane. This feature indicates the increasing amorphous nature of the polymer matrix in contact with U-2 OS cells. These results are consistent with the relationship determined by 2D-COS [34].

For the next type of nanomaterials, for PCL/ESCNFs and PCL/ESCNFs-f membranes, on the 8th day of culture, an increase in crystallinity was observed for both types of membranes. Two-dimensional correlation spectroscopy indicates the participation of carbon nanostructures in interactions with cells [33]. The increase in the I1723/I1732 intensity ratio seems to be related to the structure of nanofibers that interact with cells in a different way [38].

The relative increase in the intensities of the above-mentioned bands indicates an increase in amorphicity in the studied nanomaterials, in comparison to the reference PCL membrane for which changes almost do not happen (Figure 7). The observed trend can be correlated with the increase in the population of the cells, whose strong development in the subsequent days of the culture was monitored in fluorescence microscopy (Figure 3), which modifies the extracellular matrix and induces changes observed on the upper surface of the membranes.

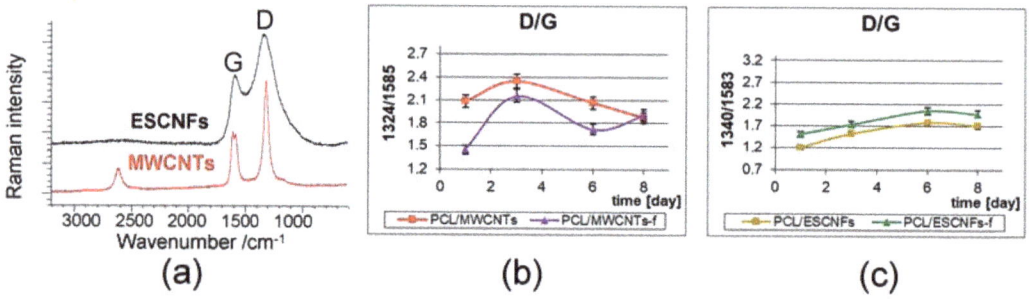

Figure 7. (a) Raman spectra of MWCNTs and CNFs nanoadditives (reference spectra), in the range 3200–500 cm^{-1}; the I(D)/I(G) intensity ratio for: (b) PCL/MWCNTs, PCL/MWCNTs-f; (Statistical analysis with PCA with Calibration 99.86237%; Validation 99.14962%) (c) PCL/ESCNFs and PCL/ESCNFs-f, (Statistical analysis with PCA with Calibration 99.86559%; Validation 99.01482%) polymer nanomaterials constituting the substrate for U-2 OS cell culture, in the following days of culture; 785 nm excitation line, variance is defined by y-axis error bars, OMNIC software.

3.4.2. The Arrangement of Carbon Nanostructures

In Figure 7a Raman spectra of carbon nanostructures, MWCNTs and ESCNFs, are shown. The Raman spectra contain, in the first order region, the G- and D-band at ca. 1590 and 1330 cm^{-1}, respectively. A characteristic parameter determining the ordering in carbon materials is the ratio of the intensity of D-band and G-band [54,56,57]. Plots reflecting the changes of this parameter on consecutive measurement days were collected for PCL/MWCNTs and PCL/MWCNTs-f and for PCL/ESCNFs and PCL/ESCNFs-f, respectively, in Figure 7b,c. For both types of carbon nanotubes in the PCL matrix, MWCNTs and MWCNTs-f, the I_D/I_G crystallinity parameter fluctuates in the first days of cell culture, reaching some stabilization and similarity after eight days. Carbon nanofibers, ESCNFs and ESCNFs-f, are characterized by a systematic increase in disorder in polymer membranes and, interestingly, achieve a similar value after eight days of cell culture, such as carbon nanotubes. This indicates a slightly different process of cell adhesion depending on the carbon nanoadditive used. It also indicates the changes taking place in the nanomaterial itself and the modifications that nanoadditives undergo during cell culture; see Figures 3 and 4.

3.4.3. Raman Spectroscopy of U-2 OS Cell Development on PCL Membranes with Fibrous Carbon Nanoparticles

U-2 OS cells have specific growth characteristics. Less than 50% of the cells are positive for collagen type I, however, positive labeling was found for molecules related to the cartilage such as collagen types II, IV, V and X [58]. The labeling profile for the U-2 OS cells remains constant and does not depend on cell density, so these osteoblastic markers, after secretion into the extracellular matrix (ECM), may be visible in the Raman spectra. In different types of collagen structures one can anticipate the presence of Gly, because this amino acid is every third residue. Actually it is monitored in the Raman spectra as visible bands of 711 and 913 cm^{-1} (Figure 5; Table 2) [44,59]. It seems convincing to pay attention to the integral intensity in the 975–930 cm^{-1} range that increases, and indicates an increase in Proline content in the ECM (Figure 8) [45] Building collagen: proline and hydroxyproline are its essential amino acid components, and can represent in some domains up to 28 and 38%, respectively [59].

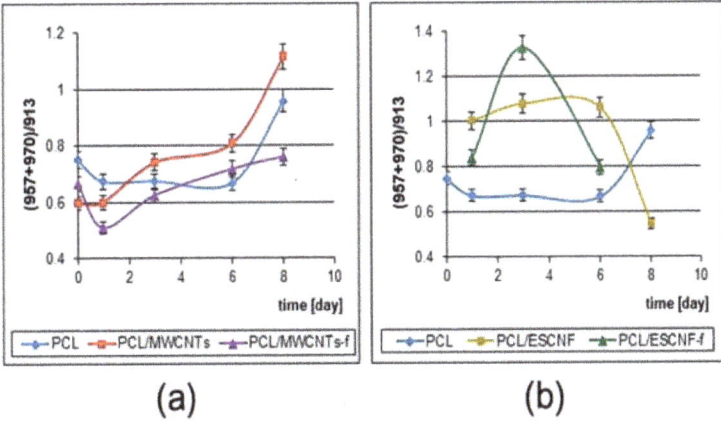

Figure 8. The intensity ratio (integral) of marker bands (957 + 970)/913 cm^{-1} characterizing the growth and development of collagen for U-2 OS cells cultured on: (**a**) PCL, PCL/MWCNTs and PCL/MWCNTs-f (Statistical analysis with PCA with Calibration 98.61861%; Validation 95.60048%); (**b**) PCL, PCL/ESCNFs and PCL/ESCNFs-f (Statistical analysis with PCA with Calibration 99.90554%; Validation 99.07312%), variance is defined by y-axis error bars, OMNIC software.

The cell growth on the studied materials is very good (Figure 3), however, the cells adhesion monitored by Raman spectroscopy proceeds in a different way, possibly due to the presence of the nanoparticle (Figure 6). The formation of the extracellular matrix may justify its influence and the observed pattern in the first days of culture, even if the U-2 OS cell line, like all osteosarcoma cell lines, shows a very heterogeneous labeling profile, which also affects the kinetics of its proliferation [58].

Types II, IV, V and X collagen show positive labeling for the U-2 OS cells line [58]. Collagen II is fibrous, the protein comprises a righthanded bundle of three parallel, left-handed polyproline II-type helices [46,59]. Type IV collagen belongs to the basement membranes and the form supramolecular networks that control cell adhesion, migration and differentiation [47]. Type V collagen is a minor component of the collagen fibrils with type I collagen [48]. The presence of type V collagen in the vicinity of the basement membranes and in the collagen fibers suggests that it can act as a linker and can also contribute to the fibril structure. Type V collagen characteristics indicate that this molecule regulates the development, differentiation and tissue repair of extracellular matrix organization [49]. The short chain of collagen X provides a pericellular matrix during ossification [50]. After secretion into the ECM, these molecules further interact to form higher supramolecular organizations that interfere with 3-dimentional nanocomposites' support. These categories of collagen include the fibrillar and network-forming proteins. They blend in very well with the structure of the nanocomposite membranes and may provide structural support for the cells and tissues.

From the second point of view the cellular metabolism products lead to progressive degradation of the membrane arrangement that is especially visible for the nanocomposite PCL/ESCNFs membrane (Figure 8b). A slightly larger number of functional groups on the carbon nanotube seems to lead to the formation of polymer matrix–MWCNTs-f nanocomposite as a tightly intertwined mat, the degradation of which is not as fast as PCL/ESCNFs. However, both these types of carbon nanotubes, MWCNTs and MWCNTs-f, and both types of carbon nanofibers, ESCNFs and ESCNFs-f, seem to very efficiently stimulate the growth of cells (Figures 4 and 8).

3.5. Morphology of U-2 OS Cells Growing on PCL Membranes with Fibrous Carbon Nanoparticles

Fluorescent images taken on the second day after seeding revealed the cells of elongated shape, which is characteristic for adherent cells growing on a cell culture surface (Figure 3) [58]. In order to further verify the initial condition of the seeded cells, at the second day of the culture the cells were stained with Hoechst 33342 fluorescent dye, which selectively binds to dsDNA molecules and stains cells nuclei. At that point, it became clear that the Hoechst 33342 stains and also the PCL revealed the hidden nanotopography of the material (Figure 9). Following the staining, the PCL micelles were clearly visible as bright areas, separated by dark grooves. A deeper look into the Hoechst-stained cell nuclei revealed no signs of necrotic disruption of the nuclei or apoptotic nuclear fragmentation/blebbing. Instead, the cells contained round-shaped, undisrupted nuclei, suggesting good condition of the cells growing on both PCL/MWCNTs materials.

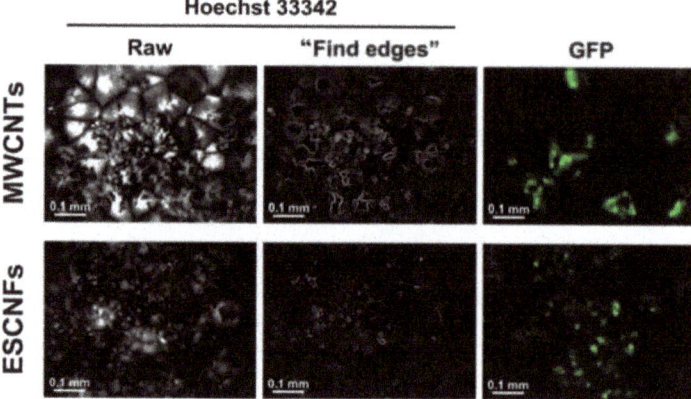

Figure 9. U-2 OS-Green cells distribution on PCL/MWCNTs and on PCL/ESCNFs on the 2nd day of culture. The specimens were live-stained with Hoechst 33342 to visualize cell nuclei. As shown in the "Raw" column, Hoechst stained not only cell nuclei but also the material (white signal, left panels). Thus, a "find edges" algorithm was used to visualize the shape of cell nuclei (the central panels). Right panels present GFP signal from the cells.

Just following the seeding (day 2), the cells tended to be located between the PCL micelles, as revealed by the Hoechst 33342 staining (Figure 9). This is presumably due to the nanotopography of the material surface, which easily allows the cells to settle in the grooves between the micelles before they adhere to their surface [10]. However, in time, while the population of the cells grew, the cells also colonized the surfaces of the micelles. This was evident after 6 days of the culture, when almost 100% of the material surface was covered by the cells (Figure 3). Interestingly, the cells were distributed almost evenly on the membrane with the second type of nanoadditive, ESCNFs. Moreover, the cells adapted to the underlying nanocomposite membranes to form the three-dimensional, thick specimens, penetrating deeper into the membrane pores. This was visible as intense, blur and scatter images of the cells (Figure 3).

4. Conclusions

Based on the fluorescence microscopy and Raman microspectroscopic data, the effect of carbon nanoadditives on the polymer structure and usefulness to stimulate the growth of bone tissue and cartilage were determined. Fluorescence microscopy that monitors the nanocomposite materials containing carbon nanoforms, PCL/MWCNTs and PCL/MWCNTs-f, as well as PCL/ESCNTs and PCL/ESCNTs-f culture with human U-2 OS cell line have shown that depending on the type of functionalization and geometric parameters of the nanoaddition, they have the bioactivity properties required for materials intended for bone tissue regeneration. The Raman spectroscopy analysis demonstrated that the degradation mechanism occurred mainly in the amorphous domains of PCL and resulted in increased polymer crystallinity, which is compatible with other reports [60]. The degradation of the membrane arrangement depends on the nanoadditive. Selected spectroscopic markers allow us to approximate the complex phenomena occurring at the interface polymer/carbon nanoaddition/U-2 OS model cell. In the current work we present research on a modified polymer with MWCNTs and we compare these results with a material modified with a completely different carbon nanoform, which is ESCNFs, i.e., a material both in terms of crystalline structure and geometric parameters different from MWCNTs. Secondly, we use an approach in which we use Raman in a way that is different from commonly applicable procedures. At the same time, we want to show which chemical changes take place in cells in contact with four types of materials and correlate them with changes occurring within nanocomposites, as well as characterize the phenomena occurring in carbon nanoforms.

Modern spectroscopic methods are therefore a significant support for other analytical methods, already in the first days of cell culture on nanomaterial. The obtained nanocomposites are promising bioactive materials for bone and cartilage tissue engineering. It should be noted that the presented materials require further assessment in accordance with applicable regulations in the context of their potential use in applications in contact with a living organism.

Author Contributions: Conceptualization, A.W.-B. and M.B.; methodology, A.W.-B. and A.K.; software, A.W.-B. and A.K.; validation, A.W.-B. and A.K.; formal analysis, A.W.-B., M.B. and A.K.; investigation, A.W.-B., A.K., M.Ś., Ł.S., E.D. and M.P.; resources, M.Ś., E.D. and Ł.S.; data curation, A.W.-B.; writing—original draft preparation, A.W.-B. and A.K.; writing—review and editing, A.W.-B., E.D., M.P. and M.B.; visualization, A.W.-B., Ł.S. and M.P.; supervision, A.W.-B. and M.B.; project administration, A.W.-B.; funding acquisition, A.W.-B. and A.K. All authors have read and agreed to the published version of the manuscript.

Funding: A.K. has been partly supported by the EU Project POWR.03.02.00-00-I004/16.

Institutional Review Board Statement: Not applicable.

Informed Consent Statement: Not applicable.

Data Availability Statement: All data generated or analyzed during this study are included in this published article. Moreover, the datasets used and/or analyzed during the current study are available from the corresponding author on reasonable request.

Conflicts of Interest: The authors declare no conflict of interest.

References

1. Daar, A.S.; Greenwood, H.L. A proposed definition of regenerative medicine. *J. Tissue Eng. Regen. Med.* **2007**, *1*, 179–184. [CrossRef] [PubMed]
2. Mason, C.; Dunnill, P. A brief definition of regenerative medicine. *Regen. Med.* **2008**, *3*, 1–5. [CrossRef] [PubMed]
3. Atala, A. Regenerative medicine strategies. *J. Pediatr. Surg.* **2012**, *47*, 17–28. [CrossRef] [PubMed]
4. Zhang, L.; Webster, T.J. Nanotechnology and nanomaterials: Promises for improved tissue regeneration. *Nano Today* **2009**, *4*, 66–80. [CrossRef]
5. O'Brien, F.J. Biomaterials & scaffolds for tissue engineering. *Mater. Today* **2011**, *14*, 88–95. [CrossRef]
6. Khang, D.; Carpenter, J.; Chun, Y.W.; Pareta, R.; Webster, T.J. Nanotechnology for regenerative medicine. *Biomed. Microdevices* **2010**, *12*, 575–587. [CrossRef]
7. Burg, K.J.L.; Porter, S.; Kellam, J.F. Biomaterial developments for bone tissue engineering. *Biomaterials* **2000**, *21*, 2347–2359. [CrossRef]
8. Roseti, L.; Parisi, V.; Petretta, M.; Cavallo, C.; Desando, G.; Bartolotti, I.; Grigolo, B. Scaffolds for Bone Tissue Engineering: State of the art and new perspectives. *Mater. Sci. Eng. C* **2017**, *78*, 1246–1262. [CrossRef]
9. Farokhi, M.; Mottaghitalab, F.; Samani, S.; Shokrgozar, M.A.; Kundu, S.C.; Reis, R.L.; Fatahi, Y.; Kaplan, D.L. Silk fibroin/hydroxyapatite composites for bone tissue engineering. *Biotechnol. Adv.* **2018**, *36*, 68–91. [CrossRef]
10. Karageorgiou, V.; Kaplan, D. Porosity of 3D biomaterial scaffolds and osteogenesis. *Biomaterials* **2005**, *26*, 5474–5491. [CrossRef]
11. Ge, Z.; Jin, Z.; Cao, T. Manufacture of degradable polymeric scaffolds for bone regeneration. *Biomed. Mater.* **2008**, *3*, 22001. [CrossRef]
12. Salinas, A.J.; Esbrit, P.; Vallet-Regí, M. A tissue engineering approach based on the use of bioceramics for bone repair. *Biomater. Sci.* **2013**, *1*, 40–51. [CrossRef]
13. Rogel, M.R.; Qiu, H.; Ameer, G.A. The role of nanocomposites in bone regeneration. *J. Mater. Chem.* **2008**, *18*, 4233–4241. [CrossRef]
14. Sheikh, Z.; Najeeb, S.; Khurshid, Z.; Verma, V.; Rashid, H.; Glogauer, M. Biodegradable materials for bone repair and tissue engineering applications. *Materials* **2015**, *8*, 5744–5794. [CrossRef]
15. Taddei, P.; Di Foggia, M.; Causa, F.; Ambrosio, L.; Fagnano, C. In vitro bioactivity of poly(ε-caprolactone)-apatite (PCL-AP) scaffolds for bone tissue engineering: The influence of the PCL/AP ratio. *Int. J. Artif. Organs* **2006**, *29*, 719–725. [CrossRef]
16. White, A.A.; Best, S.M.; Kinloch, I.A. Hydroxyapatite-carbon nanotube composites for biomedical applications: A review. *Int. J. Appl. Ceram. Technol.* **2007**, *4*, 1–13. [CrossRef]
17. Paluszkiewicz, C.; Wesełucha-Birczyńska, A.; Stodolak-Zych, E.; Hasik, M. 2D IR correlation analysis of chitosan-MMT nanocomposite system. *Vib. Spectrosc.* **2012**, *60*, 185–188. [CrossRef]
18. Abedalwafa, M.; Wang, F.; Wang, L.; Li, C. Biodegradable poly-epsilon-caprolactone (PCL) for tissue engineering applications: A review. *Rev. Adv. Mater. Sci.* **2013**, *34*, 123–140.
19. Tyan, H.L.; Wu, C.Y.; Wei, K.H. Effect of montmorillonite on thermal and moisture absorption properties of polyimide of different chemical structures. *J. Appl. Polym. Sci.* **2001**, *81*, 1742–1747. [CrossRef]
20. Wesełucha-Birczyńska, A.; Świętek, M.; Sołtysiak, E.; Galiński, P.; Płachta, Ł.; Piekara, K.; Błazewicz, M. Raman spectroscopy and the material study of nanocomposite membranes from poly(ε-caprolactone) with biocompatibility testing in osteoblast-like cells. *Analyst* **2015**, *140*, 2311–2320. [CrossRef]
21. Wesełucha-Birczyńska, A.; Morajka, K.; Stodolak-Zych, E.; Długoń, E.; Dużyja, M.; Lis, T.; Gubernat, M.; Ziąbka, M.; Błażewicz, M. Raman studies of the interactions of fibrous carbon nanomaterials with albumin. *Spectrochim. Acta Part A Mol. Biomol. Spectrosc.* **2018**, *196*, 262–267. [CrossRef]
22. Wesełucha-Birczyńska, A.; Stodolak-Zych, E.; Piś, W.; Długoń, E.; Benko, A.; Błażewicz, M. A model of adsorption of albumin on the implant surface titanium and titanium modified carbon coatings (MWCNT-EPD). 2D correlation analysis. *J. Mol. Struct.* **2016**, *1124*, 61–70. [CrossRef]
23. Tran, P.A.; Zhang, L.; Webster, T.J. Carbon nanofibers and carbon nanotubes in regenerative medicine. *Adv. Drug Deliv. Rev.* **2009**, *61*, 1097–1114. [CrossRef]
24. Fraczek-Szczypta, A.; Dlugon, E.; Weselucha-Birczynska, A.; Nocun, M.; Blazewicz, M. Multi walled carbon nanotubes deposited on metal substrate using EPD technique. A spectroscopic study. *J. Mol. Struct.* **2013**, *1040*, 238–245. [CrossRef]
25. Wesełucha-Birczyńska, A.; Frączek-Szczypta, A.; Długoń, E.; Paciorek, K.; Bajowska, A.; Kościelna, A.; Błażewicz, M. Application of Raman spectroscopy to study of the polymer foams modified in the volume and on the surface by carbon nanotubes. *Vib. Spectrosc.* **2014**, *72*, 50–56. [CrossRef]
26. Kołodziej, A.; Długoń, E.; Świętek, M.; Ziąbka, M.; Dawiec, E.; Gubernat, M.; Michalec, M.; Wesełucha-Birczyńska, A. A Raman Spectroscopic Analysis of Polymer Membranes with Graphene Oxide and Reduced Graphene Oxide. *J. Compos. Sci.* **2021**, *5*, 20. [CrossRef]

27. Hopley, E.L.; Salmasi, S.; Kalaskar, D.M.; Seifalian, A.M. Carbon nanotubes leading the way forward in new generation 3D tissue engineering. *Biotechnol. Adv.* **2014**, *32*, 1000–1014. [CrossRef]
28. Khatri, Z.; Nakashima, R.; Mayakrishnan, G.; Lee, K.H.; Park, Y.H.; Wei, K.; Kim, I.S. Preparation and characterization of electrospun poly(ε-caprolactone)-poly(l-lactic acid) nanofiber tubes. *J. Mater. Sci.* **2013**, *48*, 3659–3664. [CrossRef]
29. Kriparamanan, R.; Aswath, P.; Zhou, A.; Tang, L.; Nguyen, K.T. Nanotopography: Cellular responses to nanostructured materials. *J. Nanosci. Nanotechnol.* **2006**, *6*, 1905–1919. [CrossRef]
30. Wesełucha-Birczyńska, A.; Moskal, P.; Dużyja, M.; Stodolak-Zych, E.; Długoń, E.; Kluska, S.; Sacharz, J.; Błażewicz, M. 2D correlation Raman spectroscopy of model micro- and nano-carbon layers in interactions with albumin, human and animal. *J. Mol. Struct.* **2018**, *1171*, 587–593. [CrossRef]
31. Wesełucha-Birczyńska, A.; Stodolak-Zych, E.; Turrell, S.; Cios, F.; Krzuś, M.; Długoń, E.; Benko, A.; Niemiec, W.; Błażewicz, M. Vibrational spectroscopic analysis of a metal/carbon nanotube coating interface and the effect of its interaction with albumin. *Vib. Spectrosc.* **2016**, *85*, 185–195. [CrossRef]
32. Kołodziej, A.; Wesełucha-Birczyńska, A.; Moskal, P.; Stodolak-zych, E.; Dużyja, M.; Długoń, E.; Sacharz, J.; Błażewicz, M. 2D-Raman Correlation Spectroscopy as a Method to Recognize of the Interaction at the Interface of Carbon Layer and Albumin. *J. Autom. Mob. Robot. Intell. Syst.* **2019**, *13*. [CrossRef]
33. Wesełucha-Birczyńska, A.; Kołodziej, A.; Świętek, M.; Moskal, P.; Skalniak, Ł.; Długoń, E.; Błażewicz, M. Does 2D correlation Raman spectroscopy distinguish polymer nanomaterials due to the nanoaddition? *J. Mol. Struct.* **2020**, *1217*, 128342. [CrossRef]
34. Kołodziej, A.; Wesełucha-Birczyńska, A.; Świętek, M.; Skalniak, Ł.; Błażewicz, M. A 2D-Raman correlation spectroscopy study of the interaction of the polymer nanocomposites with carbon nanotubes and human osteoblast-like cells interface. *J. Mol. Struct.* **2020**, *1212*. [CrossRef]
35. Zaera, F. Probing liquid/solid interfaces at the molecular level. *Chem. Rev.* **2012**, *112*, 2920–2986. [CrossRef]
36. Benko, A.; Przekora, A.; Wesełucha-Birczyńska, A.; Nocuń, M.; Ginalska, G.; Błażewicz, M. Fabrication of multi-walled carbon nanotube layers with selected properties via electrophoretic deposition: Physicochemical and biological characterization. *Appl. Phys. A Mater. Sci. Process.* **2016**, *122*. [CrossRef]
37. Musiol, P.; Szatkowski, P.; Gubernat, M.; Weselucha-Birczynska, A.; Blazewicz, S. Comparative study of the structure and microstructure of PAN-based nano- and micro-carbon fibers. *Ceram. Int.* **2016**, *42*, 11603–11610. [CrossRef]
38. Panek, A.; Frączek-Szczypta, A.; Długoń, E.; Nocuń, M.; Paluszkiewicz, C.; Błażewicz, M. Genotoxicity study of carbon nanoforms using a comet assay. *Acta Phys. Pol. A* **2018**, *133*, 306–308. [CrossRef]
39. Niforou, K.N.; Anagnostopoulos, A.K.; Vougas, K.; Kittas, C.; Gorgoulis, V.G.; Tsangaris, G.T. The proteome profile of the human osteosarcoma U_2OS cell line. *Cancer Genom. Proteom.* **2008**, *5*, 63–77.
40. Schneider, C.A.; Rasband, W.S.; Eliceiri, K.W. NIH Image to ImageJ: 25 years of image analysis. *Nat. Methods* **2012**, *9*, 671–675. [CrossRef]
41. Kołodziej, A.; Wesełucha-Birczyńska, A.; Świętek, M.; Skalniak, Ł.; Błażewicz, M. Raman microspectroscopic investigations of polymer nanocomposites: Evaluation of physical and biophysical properties. *Int. J. Polym. Mater. Polym. Biomater.* **2019**, *68*, 44–52. [CrossRef]
42. Waters, J.C.; Wittmann, T. Concepts in Quantitative Fluorescence Microscopy. In *Methods in Cell Biology*; Wilson, L., Tran, P., Eds.; Elsevier Inc.: Amsterdam, The Netherlands, 2014; ISBN 0091679X.
43. Kister, G.; Cassanas, G.; Bergounhon, M.; Hoarau, D.; Vert, M. Structural characterization and hydrolytic degradation of solid copolymers of D, L-lactide-co-ε-caprolactone by Raman spectroscopy. *Polymer* **2000**, *41*, 925–932. [CrossRef]
44. Kumar, S.; Rai, A.K.; Singh, V.B.; Rai, S.B. Vibrational spectrum of glycine molecule. *Spectrochim. Acta Part A Mol. Biomol. Spectrosc.* **2005**, *61*, 2741–2746. [CrossRef]
45. Rippon, W.B.; Koenig, J.L.; Walton, A.G. Raman Spectroscopy of Proline Oligomers and Poly-L-proline. *J. Am. Chem. Soc.* **1970**, *92*, 7455–7459. [CrossRef]
46. Gelse, K.; Pöschl, E.; Aigner, T. Collagens-Structure, function, and biosynthesis. *Adv. Drug Deliv. Rev.* **2003**, *55*, 1531–1546. [CrossRef]
47. Khoshnoodi, J.; Pedchenko, V.; Hudson, B.G. Mammalian collagen IV. *Microsc. Res. Tech.* **2008**, *71*, 357–370. [CrossRef]
48. Mizuno, K.; Adachi, E.; Imamura, Y.; Katsumata, O.; Hayashi, T. The fibril structure of type V collagen triple-helical domain. *Micron* **2001**, *32*, 317–323. [CrossRef]
49. Bächinger, H.P.; Mizuno, K.; Vranka, J.A.; Boudko, S.P. Collagen Formation and Structure. In *Comprehensive Natural Products II: Chemistry and Biology*; Elsevier Ltd.: Amsterdam, The Netherlands, 2010; Volume 5, ISBN 9780080453828.
50. Bogin, O.; Kvansakul, M.; Rom, E.; Singer, J.; Yayon, A.; Hohenester, E. Insight into Schmid metaphyseal chondrodysplasia from the crystal structure of the collagen X NC1 domain trimer. *Structure* **2002**, *10*, 165–173. [CrossRef]
51. *Tu Raman Spectroscopy in Biology: Principles and Applications*; John Wiley & Sons, Ltd: New York, NY, USA, 1982.
52. Zhu, G.; Zhu, X.; Fan, Q.; Wan, X. Raman spectra of amino acids and their aqueous solutions. *Spectrochim. Acta Part A Mol. Biomol. Spectrosc.* **2011**, *78*, 1187–1195. [CrossRef]
53. Dresselhaus, M.S.; Dresselhaus, G.; Saito, R.; Jorio, A. Raman spectroscopy of carbon nanotubes. *Phys. Rep.* **2005**, *409*, 47–99. [CrossRef]
54. Wesełucha-Birczyńska, A.; Babeł, K.; Jurewicz, K. Carbonaceous materials for hydrogen storage investigated by 2D Raman correlation spectroscopy. *Vib. Spectrosc.* **2012**, *60*, 206–211. [CrossRef]

55. Rehman, I.U.; Movasaghi, Z.; Rehman, S. *Vibrational Spectroscopy for Tissue Analysis*, 1st ed.; CRC Press: Boca Raton, FL, USA, 2012; ISBN 9780429106118. [CrossRef]
56. Tuinstra, F.; Koenig, J.L. Raman Spectrum of Graphite. *J. Chem. Phys.* **1970**, *53*, 1126–1130. [CrossRef]
57. Wesełucha-Birczyńska, A.; Długoń, E.; Kołodziej, A.; Bilska, A.; Sacharz, J.; Błażewicz, M. Multi-wavelength Raman microspectroscopic studies of modified monwoven carbon scaffolds for tissue engineering applications. *J. Mol. Struct.* **2020**, *1220*, 128665. [CrossRef]
58. Pautke, C.; Schieker, M.; Tischer, T.; Kolk, A.; Neth, P.; Mutschler, W.; Milz, S. Characterization of Osteosarcoma Cell Lines MG-63, Saos-2 and U-2 OS in Comparison to Human Osteoblasts. *Anticancer Res.* **2004**, *24*, 3743–3748.
59. Shoulders, M.D.; Raines, R.T. Collagen structure and stability. *Annu. Rev. Biochem.* **2009**, *78*, 929–958. [CrossRef]
60. Woodruff, M.A.; Hutmacher, D.W. The return of a forgotten polymer—Polycaprolactone in the 21st century. *Prog. Polym. Sci.* **2010**, *35*, 1217–1256. [CrossRef]

Article

Nanostructural Arrangements and Surface Morphology on Ureasil-Polyether Films Loaded with Dexamethasone Acetate

João Augusto Oshiro-Junior [1,2], Angelo Lusuardi [1], Elena M. Beamud [1], Leila Aparecida Chiavacci [3] and M. Teresa Cuberes [1,*]

[1] Department of Applied Mechanics and Project Engineering, Mining and Industrial Engineering School of Almaden, University of Castilla-La Mancha, Plaza Manuel Meca 1, 13400 Almadén, Spain; joaooshiro@yahoo.com.br (J.A.O.-J.); angelo.lusuardi@gmail.com (A.L.); elenamaria.beamud@uclm.es (E.M.B.)

[2] Laboratory of Development and Characterization of Pharmaceutical Products, Department of Pharmacy, Center for Biological and Health Sciences, State University of Paraíba (UEPB), Campina Grande, Paraíba 58429-600, Brazil

[3] Department of Drugs and Medicines, School of Pharmaceutical Sciences, São Paulo State University (UNESP), Highway Araraquara-Jaú, Araraquara 14800-903, Brazil; leila.chiavacci@unesp.br

* Correspondence: teresa.cuberes@uclm.es

Citation: Oshiro, J.A., Jr.; Lusuardi, A.; Beamud, E.M.; Chiavacci, L.A.; Cuberes, M.T. Nanostructural Arrangements and Surface Morphology on Ureasil-Polyether Films Loaded with Dexamethasone Acetate. *Nanomaterials* **2021**, *11*, 1362. https://doi.org/10.3390/nano11061362

Academic Editor: Arthur P Baddorf

Received: 22 March 2021
Accepted: 15 May 2021
Published: 21 May 2021

Publisher's Note: MDPI stays neutral with regard to jurisdictional claims in published maps and institutional affiliations.

Copyright: © 2021 by the authors. Licensee MDPI, Basel, Switzerland. This article is an open access article distributed under the terms and conditions of the Creative Commons Attribution (CC BY) license (https://creativecommons.org/licenses/by/4.0/).

Abstract: Ureasil-Poly(ethylene oxide) (u-PEO500) and ureasil-Poly(propylene oxide) (u-PPO400) films, unloaded and loaded with dexamethasone acetate (DMA), have been investigated by carrying out atomic force microscopy (AFM), ultrasonic force microscopy (UFM), contact-angle, and drug release experiments. In addition, X-ray diffraction, small angle X-ray scattering, and infrared spectroscopy have provided essential information to understand the films' structural organization. Our results reveal that while in u-PEO500 DMA occupies sites near the ether oxygen and remains absent from the film surface, in u-PPO400 new crystalline phases are formed when DMA is loaded, which show up as ~30–100 nm in diameter rounded clusters aligned along a well-defined direction, presumably related to the one defined by the characteristic polymer ropes distinguished on the surface of the unloaded u-PPO film; occasionally, larger needle-shaped DMA crystals are also observed. UFM reveals that in the unloaded u-PPO matrix the polymer ropes are made up of strands, which in turn consist of aligned ~180 nm in diameter stiffer rounded clusters possibly formed by siloxane-node aggregates; the new crystalline phases may grow in-between the strands when the drug is loaded. The results illustrate the potential of AFM-based procedures, in combination with additional physico-chemical techniques, to picture the nanostructural arrangements in polymer matrices intended for drug delivery.

Keywords: organic-inorganic hybrid films; atomic force microscopy; ultrasonic force microscopy; sol-gel

1. Introduction

Ureasil-polyether hybrid films provide an extremely versatile matrix—platform for many different applications, including controlled drug delivery [1–3]. These materials consist of polyether macromers, such as Poly(ethylene oxide) (u-PEO) or Poly(propylene oxide) (u-PPO), linked by urea bridges to a silicate backbone. They can be prepared using the sol-gel method, with superb processability, mechanical, thermal, and chemical stability, luminescence, and biocompatibility. Their properties can be tailored using polyether macromers of different molecular weight [4], or blends of them [5].

In this study, we have prepared u-PEO and u-PPO films from polyether macromers of relatively small molecular weight (500 and 400, respectively), loaded them with dexamethasone acetate (DMA), and performed drug release experiments, with the aim of obtaining information about the conformation of their networks and how it is modified when the drug is incorporated and released. To this purpose, a series of physico-chemical techniques [6] such as X-ray diffraction (XRD), small-angle X-ray scattering (SAXS), and Fourier-transformed infrared spectroscopy (FT-IR) have been applied to the films' characterization. Contact-angle studies have also been performed, and particular attention has been devoted to the investigation of the films using atomic force microscopy (AFM)-based procedures, including ultrasonic force microscopy (UFM), which is a relatively new technique, extremely powerful for mapping surface and subsurface stiffness inhomogeneities [7–9].

The incorporation and release of several drugs from ureasil-polyether matrices has already been reported in the literature. Sodium diclofenac was incorporated into u-PEO hybrid matrices [10]. The release profile of antitumor cisplatin molecules and cisplatin-derived species from ureasil-polyether matrices have been investigated [11,12]. Triamcinolone release has also been analyzed [13]. Recently, release studies from human intragenic antimicrobial peptides from ureasil-polyether matrices were carried out [14].

To the best of our knowledge, the incorporation and release of dexamethasone acetate from ureasil-polyether matrices has not yet been studied, although it has been studied from other matrices [15–18]. Dexamethasone is an anti-inflammatory steroid drug, the utility of which in treating COVID-19 disease, in addition to vaccines, is currently under investigation [19].

The research conducted has provided us with a wealth of information on the interaction of DMA within ureasil-polyether matrices. The interrelation of data from different techniques allows for a better understanding of the nanostructural organization within organic-inorganic hybrid matrices, providing invaluable insight into aspects that may influence sustained-release technology.

2. Materials and Methods

2.1. Preparation of the Ureasil-Polyether Hybrid Materials

The ureasil-polyether hybrid materials were synthesized by the well-known sol-gel process. Briefly, a precursor was prepared from a functionalized polyether, based on Poly(ethylene oxide) (NH_2-PEO-NH_2) of molecular weight 500 g·mol^{-1} (for u-PEO) and based on Poly(propylene oxide) (NH_2-PPO-NH_2) of molecular weight 400 g·mol^{-1} (for u-PPO) dissolved in ethanol [1,2]. A modified alkoxide, 3-(isocyanatopropyl)-triethoxysilane (IsoTrEOS) (Sigma-Aldrich, São Paolo, Brasil 95% purity, CAS #24801-88-5) in a polymer/alkoxide molar ratio of 1:2 was added to this solution, and the resulting solution was maintained at reflux for 24 h at 60 °C to promote the formation of the hybrid precursor $(EtO)_3Si(CH_2)_3NHC(=O)NHCHCH_3CH_2$-(polyether)-$CH_2CH_3CHNH(O=)NHC(CH_2)_3$-$Si(OEt)_3$. Subsequently, the solvent was removed using a rotary evaporator (IKA RV 10, Staufen, Germany) operated at 60 °C and 175 mbar.

To prepare the films, the precursor was dissolved in water and ethanol in an appropriate vessel, and HCl was added as a catalyst to subject the precursor to the sol-gel hydrolysis and condensation reactions, in the proportion 500 µL ethanol, 25 µL water, and 25 µL HCl catalyst to 0.75 mg of ureasil-polyether hybrid precursor. To load the drug, crystalline DMA powder (dexametasona acetate micro, SM Empreendimentos Farmaceuticos Lta São Paolo, Brasil, CAS: 1177-87-3) in 3% wt/wt proportion to the precursor in our case was dissolved in the ethanol/water solution, and the precursor and then the HCl catalyst were added to induce the reactions. Films of 1 mm thickness were typically prepared.

2.2. In Vitro Drug Released

The u-PEO500 and u-PPO400 films were immersed in 900 mL of medium (phosphate buffer 7.2 pH with 0.5% of procetyl AWS® (CRODA, Rawcliffe Bridge, UK)) to guarantee the sink condition at 37 ± 0.5 °C and were stirred with a USP dissolution apparatus 2 (paddle) at a speed of 50 rpm. At time intervals, 5 mL of filtered release medium was removed for analysis and replaced with the same volume of medium. The DMA amount in the extracted solutions was analyzed by UV-vis absorbance at 241 nm, using a UV-Vis Cary 60 Spectrophotometer (Agilent Technologies, Melbourne, Australia). The cumulative percentage of drug release was calculated from the average of three parallel monitoring.

The results were expressed as the mean ± SD of three experiments. All results obtained in the in vitro drug release study are presented as means and standard deviations (SD). The results were compared by ANOVA and post-hoc Tukey. The significance level (p) adopted was 0.05. Statistical analyses were performed with the program Instat for Windows (GraphPads software, San Diego, USA). Drug release kinetics was analyzed by plotting the mean release data versus time, which were fitted with different mathematical models [20]. In all cases the SigmaPlot 10.0 program (Systat Software Inc, San Jose, CA, USA) was used.

2.3. X-Ray Difraction

X-ray Diffraction measurements were performed in an equipment Philips X'Pert MPD (Eindhoven, Holland), using CuK α radiation (1.54056 Å) with 40 KV and 40 mA. It incorporates 0.04 rad soller slits for both incident and diffracted beams, an automatic 12.5 mm programmable divergence slit, and a Xe gas sealed proportional detector. Data were collected in an angular range between 1° and 50° (2θ) with a step size of 0.01° and a counting time of 0.70 s per step. The data analysis was carried out with the Fityk software (Varsaw, Poland) [21] (open source).

2.4. Small Angle X-ray Scattering

SAXS measurements were performed at the NCD beamline of ALBA Synchrotron (Barcelona, Spain). The beamline was equipped with a 2D Pilatus 300 k detector located 910.9 mm from the sample, recording the image of the scattering intensity, $I(q)$, as a function of the modulus of the scattering vector, $q = 4\pi/(\lambda \sin(\frac{\varepsilon}{2}))$, where ε is the X-ray scattering angle. The data were normalized considering the varying intensity of the direct X-ray beam, the detector sensitivity, and the sample transmission. The GSAS-II software (Argonne, IL, USA) [22] (open source) was used.

2.5. Fourrier-Transformed Infrared Spectroscopy

FTIR spectra (4 cm^{-1} resolution, wavenumber range 500–4000 cm^{-1}) were recorded using a Shimadzu IRPrestige-21 spectrometer (Tokio, Japan), using the ATR method. Small pieces of the ureasil-polyether hybrid films (≈1 mm thick) were cut and placed in the instrument sample holder. The data were acquired and analyzed using the software Shimadzu IR solution 1.21 (Tokio, Japan).

2.6. Scanning Probe Microscopy

Contact-mode atomic force microscopy (AFM), lateral force microscopy (LFM) and ultrasonic force microscopy (UFM) were performed using Brucker Multimode III (Santa Barbara, CA) (AFM/LFM) and NANOTEC (Madrid, Spain) (AFM/LFM and UFM [7]) instruments). For UFM, ultrasonic frequencies of ~3.8 MHz and modulation frequencies of 2.4 KHz were applied from a piezoelectric element placed under the sample. Typically, Olympus Silicon Nitride cantilevers with a nominal spring constant of 0.06 N/m and a nominal tip radius of 20 nm were used. The measurements were performed in air, at ambient conditions. Data analysis was performed with WSxM software (Madrid, Spain) [23].

2.7. Contact-Angle (Wettability)

Wettability tests (hydrophobic/hydrophilic) were performed based on the contact angle of water droplets on ureasil-polyether hybrid films. Droplets of 10 µL water were applied at a rate of 2 µL/s, using a 15+ OCA (Dataphysics) apparatus and SCA software 20.2.0 (DataPhysics Instruments GmbH, Filderstadt, Germany), evaluating the measurements after 15 s. Film samples ~2 cm in diameter and ~1 mm thick were used, and the experiments were carried out at room temperature.

3. Results and Discussion

Figure 1 shows the structural formulas of the precursor molecules of u-PEO and u-PPO, and DMA. For u-PEO500, the PEO chain length contains less than n = 12 oxyethylene units, and for u-PPO400, the PPO chain length contains less than n = 7 oxypolypropylene units. As hydrolysis and condensation reactions take place during the sol-gel process, the silanol terminal groups of different molecules interact with each other to form the inorganic siloxane nodes that create the matrix network. Besides, interactions among the urea and polyether moieties from different molecules may also occur. In particular, in low molecular weight ureasil-polyether like ours, the number of urea-urea linkages is expected to be quite large [4]. The formation of hydrogen-bonded urea-polyether associations is also possible, as the N-H groups of the urethane linkages are donor sites, and the ether oxygens, hydrogen bond acceptors. As the DMA molecules dissolve together with the precursor molecules when preparing the films, they may also interact and/or influence the sol-gel reactions leading to film formation [13].

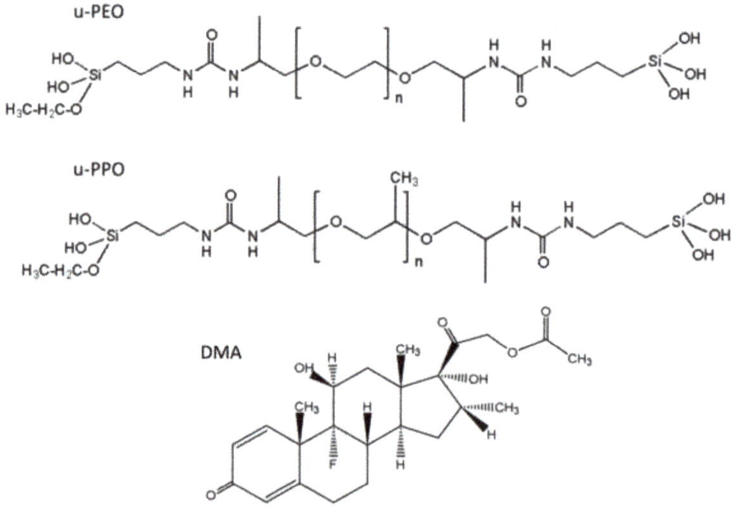

Figure 1. Structural formulas of u-PEO, u-PPO, and dexamethasone acetate.

Figure 2 presents the release profiles of DMA-loaded u-PEO500 and u-PPO400 films. Figure 2a allows us to visualize the different release rates of the films. Figure 2b,c show the different mathematical models applied to fit the DMA release profile in each case.

As it is clearly noticeable from Figure 2a, the samples prepared with u-PEO500 exhibit much faster release rates than those prepared with u-PPO400. Typically, release from u-PEO matrices is much faster than from u-PPO matrices, because in hydrophilic u-PEO the drug molecules can easily diffuse into the release medium through the free volume of the swollen network [1,2]. Whereas u-PEO has a highly hydrophilic character, the presence of the additional methyl group in u-PPO decreases its hydrophilicity. As a result, its affinity for the dissolution medium is decreased, and a lower relaxation of the polymer chains and a lower degree of swelling of the u-PPO matrix is expected. In our case, u-PEO500 and u-PPO400 hybrid materials have similar molecular weight, so the molecular weight of the polymer chain is not a factor in determining the different release profiles.

Fitting drug release data using mathematical models provides a tool to elucidate the main transport mechanisms that control the drug release process [20,24–27]. In Figure 2b,c Higuchi [28], Peppas [29] and Weibull [30] models have been considered. The criterion used to choose among these models is the statistical coefficient of determination (r^2), which is used to evaluate the fit of the model equation. According to the values of r^2, the DMA release from u-PEO500 (Figure 2b) fits best with the Peppas model, while that of u-PPO400 (Figure 2c) fits best with the Weibull model.

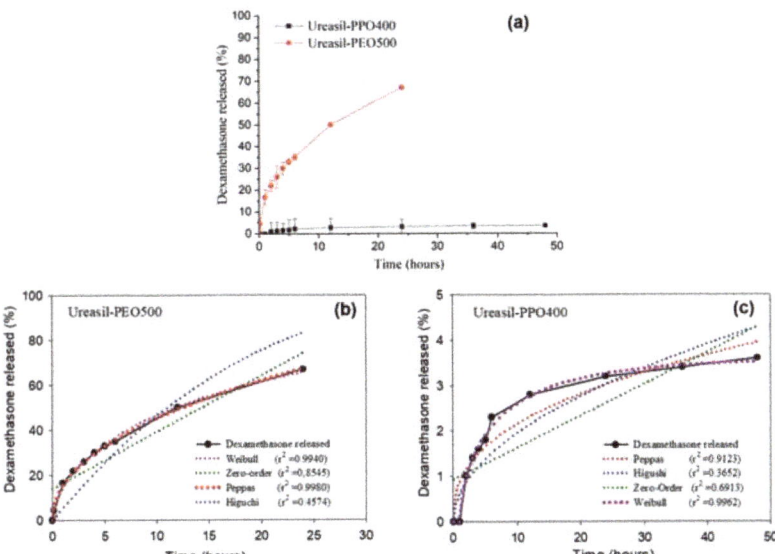

Figure 2. (a) In vitro dissolution profiles of DMA from u-PEO500 and u-PPO400 films. (b,c) Fittings of different mathematical models to the DMA release profile from (b) u-PEO500 and (c) u-PPO400.

The Peppas model is based on a power law correlation between drug release and time (($release\ of\ DMA) = constant \cdot (time)^n$). For the Peppas model the values of exponent n determine the mechanism of drug transport out of matrix [29]. When values of n are less than 0.45, release is expected to occur by Fickian diffusion, whereas values of n between $0.45 < n < 0.89$ suggest that the release is governed by anomalous transport, involving both matrix swelling and drug diffusion. In a swelling-controlled release mechanism, release depends mostly on solvent penetration. For u-PEO500, the value of the exponent n is 0.4574, indicating that the release of the drug from the matrix to the medium presumably occurs by anomalous transport.

The Weibull model is an empirical approach, not based in any kinetic theory (($release\ of\ DMA) = 1 - \exp[-constant \cdot (time)^b]$). Nevertheless, reports in the literature indicate that the Weibul model does provide information about the drug release process, with the value of the exponent b correlated with the mechanism of drug transport out of the matrix [30]. When the value of b is less than 0.75, release is expected to occur by Fickian diffusion. The u-PPO presented a b exponent value of 0.7409. Therefore, the release of DMA from DMA-loaded u-PPO400 presumably occurs mostly by Fickian diffusion.

Figure 3 shows the topography of (5000 × 5000) nm surface areas recorded on unloaded (a) u-POE500 and (b) u-PPO400 films using contact-mode AFM. Figure 3c,d correspond to height contour profiles along the continuous white lines in Figure 3a,b, respectively. Figure 3e,f are 3D representations of Figure 3a,b. The surface of the u-PEO500 film is characterized by scattered pores as large as ≈200 nm in diameter; some of the pores have been pointed out with dashed white circles in Figure 3a, and are apparent in Figure 3e. On the u-PPO400 films' surfaces polymer "ropes" are apparent (see Figure 3b,f) oriented along a well-defined direction, indicated by a dashed white arrow in Figure 3b. The root mean square (RMS) roughness on Figure 3a (u-PEO500) and Figure 3b (u-PPO400) is of 1.80 nm and 1.35 nm, respectively. The presence of pores causes the colour-scale range in the u-PEO image to be much larger than in the u-PPO image (see Figure 3e,f). Nevertheless, in the areas where there are no pores, the surface corrugation is only slightly larger on the u-PEO than on the u-PPO surface, as can be seen in Figure 3c,d. The pores favor the penetration of water into the matrix, leading to swelling, and thus modifying the surface morphology. The presence of pores helps to explain the higher drug release rates in u-PEO than in u-PPO films [31].

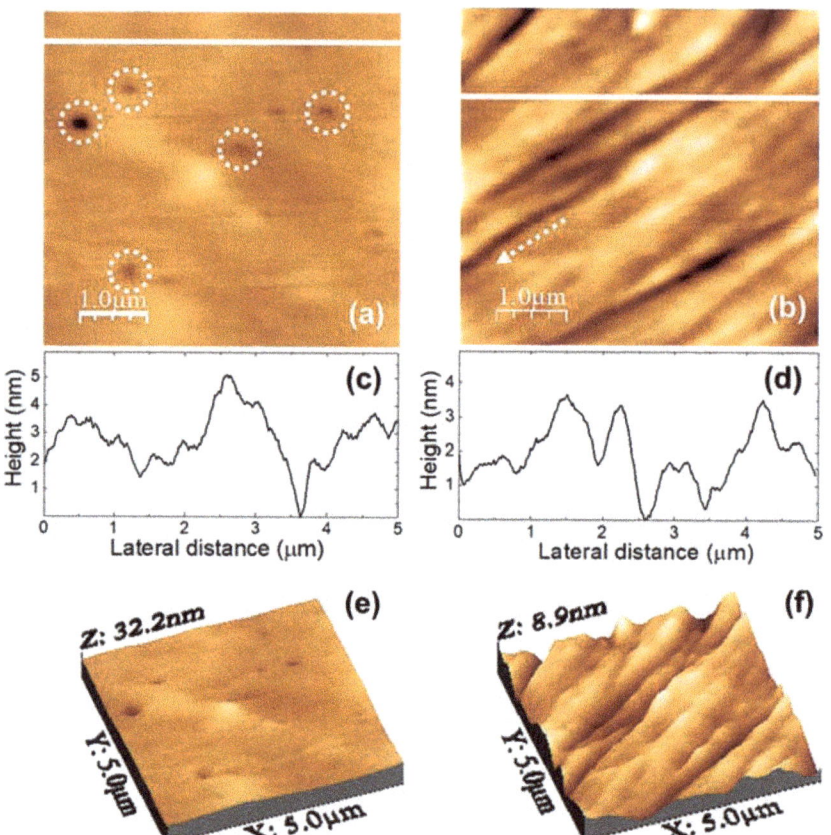

Figure 3. (**a**,**b**) Contact mode AFM topographic images of (**a**) unloaded u-PEO500 and (**b**) unloaded u-PPO400 films. (**c**,**d**) Height-contours profiles along the white lines in (**a**,**b**), respectively. (**e**,**f**) 3D representations of (**a**,**b**), respectively.

The images in Figure 4 were recorded over a (5000 × 5000) nm surface area of a DMA-loaded u-PEO film. Figure 4a is the surface topography, and Figure 4b,c lateral force microscopy (LFM) images scanning from right to left (Figure 4b, recorded simultaneously with Figure 4a) and from left to right (Figure 4c, each line recorded as the tip travelled back along the line when recording Figure 4b). Figure 4d shows a height-contour profile along the continuous white line in Figure 4a, and Figure 4e is a 3D representation of Figure 4a. The surface morphology appears slightly more compact than this of the unloaded u-PEO500 film (Figure 3a,e). In Figure 4a,e we may observe a groove, and Figure 4d shows that the areas without groove exhibit a corrugation a bit larger than those of the areas without pores on the unloaded u-PEO surface (Figure 3c). No traces of the presence of DMA can be distinguished on the film surface. The absence of frictional contrast in the LFM images Figure 4b,c confirms the chemical homogeneity of the film surface.

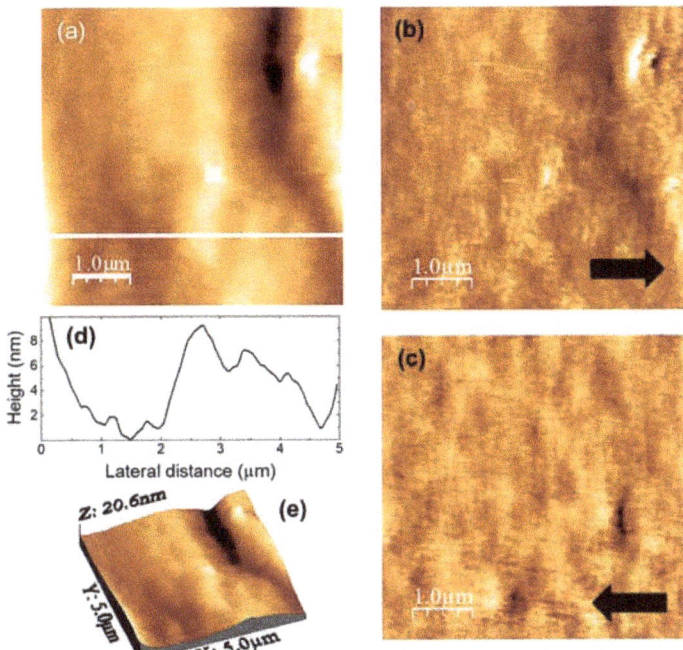

Figure 4. DMA-loaded u-PEO500. (**a**) Contact-mode AFM topography. (**b**,**c**) Lateral force microscopy (LFM) images recorded scanning (**b**) scanning from left to right (simultaneously with (**a**)) and (**c**) from right to left. (**d**) Height-contour profile along the continuous white line in (**a**). (**e**) 3D representation of (**a**).

In contrast, on DMA-loaded u-PPO films' surfaces the coexistence of different species or domains is apparent. The images in Figure 5a,b were simultaneously recorded over a (5000 × 5000) nm surface area. Figure 5a is the surface topography, and Figure 5b is the ultrasonic force microscopy (UFM) image, which allows us to distinguish nanoscale regions with different stiffness and/or adhesion [7,32]. The surface morphology in Figure 5a is markedly different from that of the unloaded u-PPO surface (Figure 3b); after loading with DMA, clusters of different sizes can be distinguished on the film surface. The largest clusters in this image, marked with dashed circles in Figure 5a,b are ~300 nm in diameter, and ~80 nm in height. In the UFM image (Figure 5b), such clusters give rise to a brighter contrast, indicative of a higher stiffness. Interestingly, in UFM, sample regions with distinctly darker contrast are clearly noticeable. Even though, in principle, darker contrast in UFM should indicate softer areas, in polymer nanocomposites, depending on the filler/matrix interface properties, the locations of the filler appear with a darker contrast in UFM due to ultrasound scattering at the interface regions [32]. The morphology in those regions (Figure 5a) is characterized by rounded clusters or dots of ~100 nm in diameter, which appear aligned along a well-defined direction, indicated by a dashed white arrow in Figure 5a. Figure 5c,d are LFM images recorded over the same surface area than Figure 5a,b, immediately after these, after suppression of the ultrasonic excitation. In this case, it is clearly noticeable that in the regions in which the UFM contrast appeared darker, such as those between the two dashed white lines outlined in Figure 5c,d, the surface dots in the LFM images exhibit different frictional contrast, with lower friction then their surroundings (LFM image contrast darker when scanning from right to left (Figure 5c), and brighter when scanning from left to right (Figure 5d)). Figure 5e shows a contact-mode topographic image recorded on a different area which also shows the tendency of the dots to align along a well-defined direction; in this image, the coalescence of some dots into segments along that direction is even clearer than in the area in Figure 5a. Figure 5f displays a height contour profile along the continuous white line in Figure 5e, showing that the dots are here smaller than ~200 nm in diameter, and 40 nm in height. Figure 5g correspond to a (25 × 25) μm topographic image recorded over a different surface area of the DMA-loaded u-PPO film. Here, the surface morphology unequivocally reveals the formation of a needle-shaped crystal within the u-PPO film, as a result of the incorporation of DMA. DMA crystallizes in the orthorhombic crystal, and the needle-like crystal in Figure 5g exhibit planes with angles of 90°, characteristics of the orthorhombic morphology. Hence, according to the information provided by the applied AFM techniques, the incorporation of DMA molecules in the u-PPO precursor solution results, after the completion of the sol-gel process, mostly, in the formation of small, rounded clusters or dots, differentiated from the matrix film, which tend to be aligned along a well-defined direction, and, also, in the occasional formation of bigger apparently crystalline needle-like clusters oriented along the same direction.

DMA has a great tendency to form solvents [33]. Two anhydrous polymorphic forms (Form I and II) are known, and two different monohydrated polymorphs have been reported (DEX I and DEX II) [34,35]. In addition, DMA presents an original behavior during its crystallization in specific conditions, and the formation of whisker-like DMA crystals that very much resemble the needle-like crystal observed in Figure 5g corresponding to a sesquihydrated form, has also been reported in the literature, for example, when saturated ethanolic solution of DMA is injected in water, or when DMA crystallized as dimethylsulfoxide (DMSO) solvate is immersed in water [36–38]. The mechanism of formation of these whiskers is not yet fully understood, but it is believed to involve a high local saturation that leads to precipitation with dendritic growth. During the sol-gel process, the solvent evaporates while water is formed during the condensation/hydrolysis reactions of the precursors' molecules, so it is plausible that the formation of the needle-like crystals in Figure 5g originates similarly to the aforementioned DMA whiskers. The observation of these crystals in u-PPO400 is of high interest [39].

DMA has previously been incorporated into bioresorbable films of poly(lactic acid) (PLA) and copolymers of lactic acid and glycolic acid (poly(DL-lactic-co-glycolic acid), (PDLGA)), and the morphology of the loaded films has been studied in detail [16,17]. The films were prepared by solution casting, with the polymer dissolved in a solvent and mixed with the drug prior to casting, followed by isothermal heat treatment after preparation. The DMA location/dispersion in the film was controlled by considering solubility effects in the starting solution and the solvent evaporation rate, which determined the kinetics of drug and polymer solidification. For high drug concentrations in the initial solution, and fast solvent evaporation rates, films with small drug particles and crystals were obtained, with drug nucleation and segregation, as well as merging of the small DMA particles to form larger crystals occurring within the dense polymer solution. In ureasil-polyether films prepared by the sol-gel procedure, the film time formation, and therefore the solvent evaporation time, is relatively fast, faster for unloaded u-PPO than for u-PEO, of the order of ~200 s for precursor molecular weights of 2000 and 1900 g·mol^{-1}, respectively [40], and the DMA molecules could behave similarly, allowing us to explain the origin of the small dots observed in DMA-loaded u-PPO films, and their apparent coalescence into segments, noticeable, for instance, in Figure 5e.

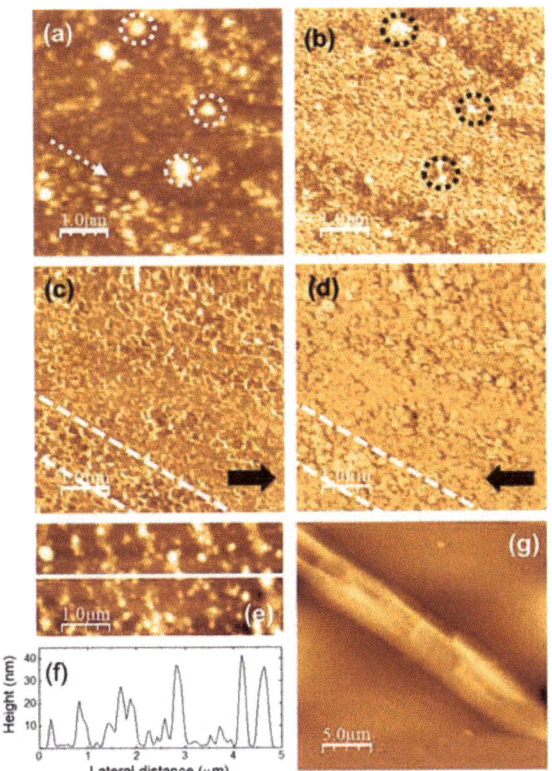

Figure 5. DMA-loaded u-PPO400. (**a**) Contact-mode AFM topography; color-scale range: 64 nm. (**b**) Ultrasonic force microscopy (UFM) image, recorded simultaneously with (**a**). (**c**,**d**) Lateral Force Microscopy (LFM) images recorded scanning (**c**) from left to right, and (**d**) from right to left. (**e**) Contact-mode AFM topography; color-scale range: 55 nm. (**f**) Height-contour profile along the continuous white line in (**e**). (**g**) Contact-mode AFM topography; color-scale range: 2 µm.

Figure 6 shows the XRD data recorded on unloaded (blue curves) and DMA-loaded (red curves) u-PEO500 and u-PPO400 films, as well as on the original DMA powders (top black curve). Both unloaded u-PEO and u-PPO films exhibit a broad peak characteristic of amorphous materials with its maximum located at 2θ = 21.6° in u-POE, and 2θ = 21.4° in u-PPO. Interestingly, an additional small peak is apparent in the unloaded u-PPO spectra, located at 2θ = 16.8°, which indicates a small amount of crystallinity. This peak corresponds to an interplanar spacing of 5.2 Å, which coincides with the 200 reflection of the orthorhombic structure of crystalline PPO [41,42]. However, for u-PPO400, the PPO moieties in the precursor molecules are quite short (see Figure 1, n < 7 for the u-PPO precursor), and not necessarily isotactic. This result is quite surprising, since to our best knowledge, even though in some cases coexistence of crystalline and amorphous phases has been observed, for instance, in u-PEO samples obtained from high molecular weight precursors (of molecular weight around 1000 g/mol or higher) [10], no trace of a crystalline phase has been previously observed in ureasil-polyether films obtained from low molecular weight precursors.

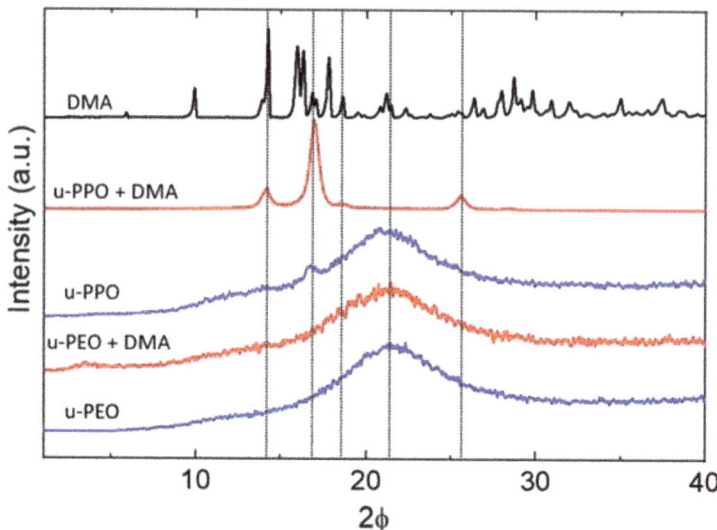

Figure 6. X-ray diffraction patterns for unloaded u-PEO and u-PPO films (blue curves), DMA-loaded u-PEO and u-PPO films (red curves) and DMA powder (black curve).

The diffractogram of the DMA-loaded u-PEO film does not experience any significant modification with respect to that of the unloaded u-PEO. In contrast, in the DMA-loaded u-PPO sample, the data clearly reveal the presence of a crystalline phase. Table 1 lists the reflections observed in the DMA-loaded u-PEO diffractogram, the corresponding interplanar spacing, and the crystallite size, derived from Debye-Scherrer equation.

Table 1. Peaks in the DMA loaded u-PPO diffractogram.

2θ/Degrees	d-Spacing (Å)	Crystallite Size (nm)
14.0094	6.3163	26.11
16.9142	5.2375	27.19
25.5421	3.4845	27.58

It is noticeable from Figure 6 that the peak at 16.8° on the unloaded u-PPO matrix is increased in the DMA-loaded film, and slightly shifted to 16.9°. It happens that this peak is coincident with a reflection from the monohydrated DMA form (DEX I) [34,36]. The peak at 14.0° corresponds to most intense peak in the original DMA powder diffractogram (top black curve in Figure 6) and can be found as a strong peak in the diffractogram of the DMA anhydrous variety (FORM II) [36], but not in this one of the monohydrated phase. Regarding the peak at 25.5°, it is coincident with the 210 reflection of the orthorhombic structure of crystalline PPO [41,42], and it is not present as a strong peak in either the anhydrous or the monohydrated DMA phases. Hence, in view of the experimental data, we cannot reject the possibility that a crystalline PPO phase forms within the u-PPO matrix, enhanced by the presence of DMA, which also appears to be present in crystalline state.

At first sight, it is surprising that the peaks of the sesquihydrate phase of DMA, which should correspond to the needle-like crystals in Figure 5g, considering its resemblance to the DMA whiskers, do not appear in the XRD diagram. We attribute this to the fact that this crystal phase is forming in small quantity, and just at the sample surface, while the information recorded in XRD comes from a surface region of several microns.

For the application of the Debye-Scherrer equation that allows us to estimate the crystallite sizes of the present phases we have considered the dimensionless shape factor K as 0.94, i.e., presuming that the domains have rounded shape. According to the morphology in Figure 5a,e this should indeed be the case. The obtained data of ~27 nm for the crystallite size is also consistent with the AFM images in Figure 5. Even though the largest clusters in Figure 5a reach up to ~300 nm, and in Figure 5e ~100 nm, there are many smaller dots of ~30 nm in size. Furthermore, the ones that we may appreciate in the image are located on the film surface, and it is to be expected that the ones located below are probably somewhat smaller, being constrained by the surrounding matrix.

Figure 7 shows the SAXS curves recorded on (a) u-PEO500 and (b) u-POP400. The blue curves correspond to unloaded samples and the red to samples loaded with DMA. The SAXS intensity, I(q), is proportional to the Fourier transform of the correlation function of the electronic density of the material. The peaks give evidence of a strong spatial correlation between the ureasil-polyether cross-linked siloxane nodes. For both unloaded u-PEO and u-PPO the maximum peak position is located at $q = 2.5$ nm^{-1}, and the averaged most probable distances between two adjacent siloxane nodes within the polymer matrix can be estimated by $d = \frac{2\pi}{q_{max}} = 2.5$ nm, where q_{max} is the modulus of the scattering vector at the peak's maximum. As seen in Figure 7, the introduction of DEXA does not significantly alter this distance, the maxima of the red curves in Figure 7a,b remain approximately in the same position as those of the blue ones. An average size of the correlation volume associated with the spatial distribution of siloxane nodes, L_c, can be obtained by applying the Scherrer equation in the case of low-angle X-ray scattering ($L_c = 4\pi/\Delta q$) where Δq is the full width at half-maximum of the correlation peak of the SAXS function [43,44]. The values of L_c for the SAXS peaks in Figure 7 are ~9.0 nm for the unloaded and DMA-loaded u-PEO films, and ~12.5 nm for the unloaded and DMA-loaded u-PPO films.

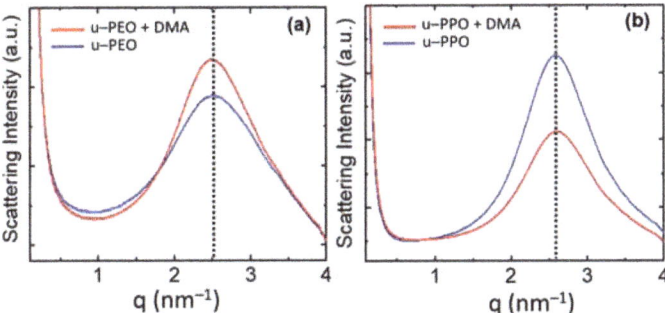

Figure 7. Experimental SAXS intensity $I(q)$ of (**a**) unloaded u-PEO 500 (blue curve) and DMA-loaded u-PEO500 (red curve) and (**b**) unloaded u-PPO (blue curve) and DMA-loaded u-PPO400 (red curve).

Interestingly, as DMA is introduced in u-PEO, the intensity of the correlation peak increases, indicating the increase of the electronic-density contrast between the ureasil nodes and the polymeric matrix, as observed in other cases [1,43]. On the other hand, in u-PPO films, the intensity decreases.

Figure 8a,b corresponds to AFM and UFM images recorded over a (2500 × 2500) nm surface area on the unloaded u-PPO film. Figure 8a is the surface topography recorded in contact-mode and Figure 8b is the UFM image recorded simultaneously with Figure 8a, over the same surface area. As Figure 3b,f, Figure 8a shows that the u-PPO400 matrix is formed by polymer ropes, oriented along a well-defined direction. The UFM image (Figure 8b) makes it possible to distinguish that the ropes are formed by "strands" ~180 nm in diameter, which in turn are structured in rounded clusters. Figure 8c is a crop of the area delimited by the white rectangle in Figure 8b, in which some rounded clusters have been enclosed by white circles to facilitate their identification. The clusters yield a brighter contrast in UFM, which indicates that they correspond to stiffer regions. The contrast in Figure 8b can be understood if the siloxane nodes within the u-PPO400 matrix assemble into "hybrid clusters", ~180 nm in diameter, corresponding to the rounded clusters seen in the image. In fact, according to the information provided by SAXS (Figure 7), each of these clusters should be formed by ~14 (i.e. (cluster diameter)/Lc) disordered hybrid "supercrystals" defined by aggregates of siloxane nodes with average separation distances of ~2.5 nm between them.

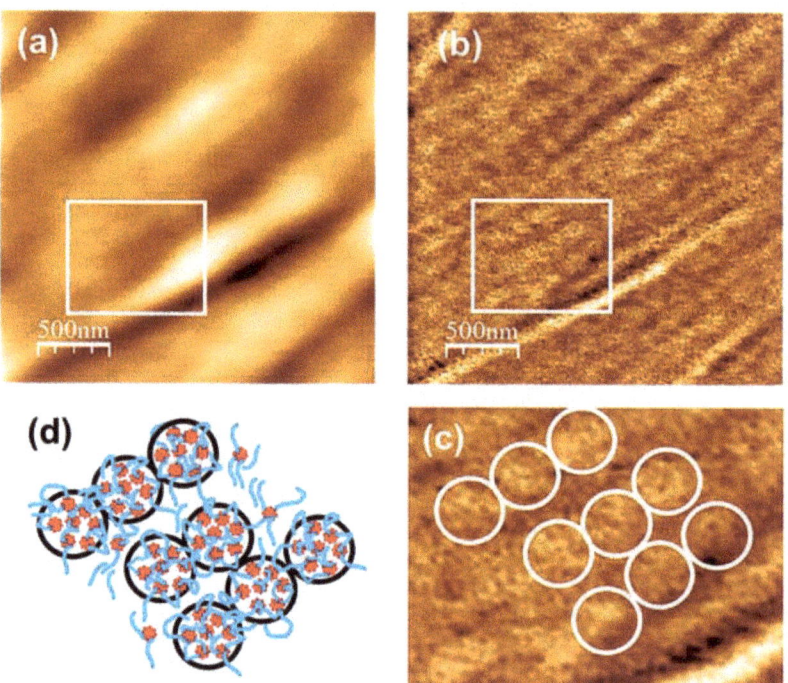

Figure 8. Unloaded u-PPO400. (**a**) Contact-mode AFM topography; color-scale range: 64 nm (**b**) Ultrasonic force microscopy (UFM) image, recorded simultaneously with (**a**); (**c**) Cropping of the part of (**b**) indicated by the white rectangle (**d**) Schematic drawing to illustrate the nanostructural arrangement within the polymer matrix.

Figure 8d depicts a tentative sketch of the u-PPO400 matrix structure, in which the red stars represent the hybrid supercrystals of siloxane nodes aggregates that give rise to the SAXS correlation peak in Figure 7b.

The model in Figure 8d provides insight into the results of Figure 5. The new crystalline phases formed within the DMA-loaded u-PPO400 matrix may occupy the sites in-between the strands defined by the aligned clusters in Figure 8d. Furthermore, the "stiffer clusters" observed on the surface of the DMA-loaded u-PPO400, (those enclosed with dashed circles in Figure 5a,b), surely correspond to some of the clusters in Figure 8b, displaced from their sites as the new phases formed upon drug loading.

Figure 9 displays the FT-IR spectra recorded on unloaded (blue curves) and DMA-loaded (red curves) (Figure 9a–c) u-PEO and (Figure 9d–f) u-PPO films. The black curves in Figure 9 correspond to the FT-IR spectrum recorded on the original DMA powders.

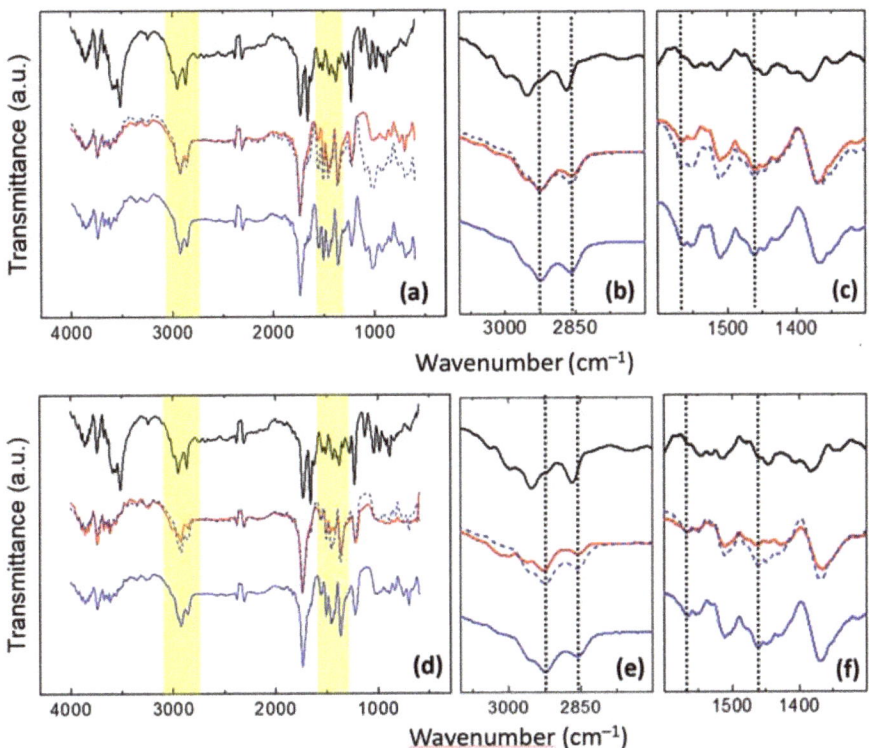

Figure 9. FT-IR spectra of (**a–c**) pure u-PEO 500 films (blue curves), DMA loaded u-PEO films (red curves) and DMA powder (black curves); and (**d–f**) pure u-PPO400 films (blue curves), DMA loaded u-PPO films (red curves) and DMA powder (black curves). In each plot, the dashed blue curve is identical to the continuous blue line and has been shifted vertically to facilitate the comparison of the unloaded and DMA-loaded films' spectra.

Table 2 lists the main vibrational peaks observed in the FT-IR spectra and their assignments [4,34,45]. Due to the similarity of the functional groups, most of the characteristic vibrational bands of the ureasil-polyether matrix and the DMA molecules occur in the same regions.

Table 2. Peak assignment in the FT-IR spectra in Figure 9.

FT-IR Wavenumber (cm^{-1})	Peak Assignment
3516	O-H stretching
3050–2850	C-H stretching
1740, 1660	C=O stretching
1565	Amide II
1475–1427	CH_2 scissoring/CH_3 deformation
1370–1351	CH_2 wagging
1100	C-O stretching

For the discussion of the information provided by the FT-IR data we will mainly focus on the yellow regions in Figure 9a,d enlarged in Figure 9b,c,e,f.

Figure 9b,e shows the C-H stretching region. Here, in both unloaded u-PEO and u-PPO curves (blue curves in Figure 9b,e), the peaks at 2925 cm^{-1} and 2859 cm^{-1}, assigned to CH_2 antisymmetric and symmetric stretching, respectively [4], have been indicated with a vertical dashed line. In the original DMA powder spectrum (black curves in Figure 9b,e), we encounter several bands in this region corresponding to olefinic and aliphatic CH stretching [45]. In DMA-loaded u-PEO films, the incorporation of DMA to the matrix does not bring much change in the FT-IR bands (see red curve in Figure 9b). Nevertheless, in DMA-loaded u-PPO films, the incorporation of DMA is accompanied by a reduction in the CH_2 stretching vibrations observed in the unloaded samples, and by the emergence of a small peak at 3020 cm^{-1} which can be related to olefinic CH stretching from the DMA molecules (see red curve in Figure 9e). The fact that the CH_2 stretching vibrations diminish confirm that the u-PPO structure is being altered in the presence of DMA, which is in agreement with the results obtained from the AFM and XRD studies.

Figure 9c,f includes the amide and CH_2/CH_3 absorption bands. The peak at 1564 cm^{-1} indicated by a vertical dashed line, discernible in both spectra of the unloaded u-PEO and u-PPO films (see blue curves in Figure 9c,f), has been previously assigned to the amide II band [4]. The amide II mode is a mixed contribution of the N-H in-plane bending, the C-N stretching and the C-C stretching vibrations, and it is very sensitive to both chain conformation and intermolecular hydrogen bonding. As it is well known, the urea compounds have a very strong self-association capacity, and in small molecular weight ureasil-polyether, the formation of many urea-urea linkages is expected. The N-H groups of the urea moieties are prone to form strong hydrogen bonds with either the carbonyls of the urea moieties of another molecules, and/or with the ether oxygens of the polyether moieties, capable to act as hydrogen bond acceptors. Interestingly, in DMA-loaded u-PEO film this peak strongly diminishes (Figure 9c), indicating that the presence of DMA in u-PEO alters the interactions of the N-H that exist in the unloaded matrix.

Consistent with our previous observation, the peak at 1100 cm^{-1} in the u-PEO film (Figure 9a, blue curve) which falls into the COC stretching region [46], strongly diminishes when DMA is incorporated (Figure 9a, red curve). In contrast, the amide II band does not experience any changes in u-PPO films when they are loaded with DMA (Figure 9f). In u-PPO, due to the presence of the additional CH_3 group, it is conceivable that the DMA molecules cannot easily access the ether oxygens of the polypropylene moieties. DMA may interact with the u-PPO matrix at sites near the ureasil nodes but if the DMA molecules reach the sites amidst the strands of clusters in Figure 8, where there are no ureasil nodes, they are bound to interact with each other and form the new phases observed by both XRD and AFM. The fact that u-PPO is structured in strands formed by aligned clusters which consist in aggregates of siloxane-nodes, as illustrated in Figure 8, explains that the new DMA phases form along a well-oriented direction, growing between the cluster strands.

The peak at 1460 cm^{-1}, also marked with a vertical dashed line and noticeable in both spectra of the unloaded u-PEO and u-PPO films (blue curves in Figure 9c,f) is assigned to a C-H mode. Consistent with the results discussed in relation to Figure 9b,e, this band experience no changes when u-PEO is loaded with DMA, but it is severely affected (the intensity of the band is reduced, i.e., the vibrational modes are hindered) when DMA is incorporated into the u-PPO films.

Figure 10 presents the values obtained from contact-angle measurements (θ) on the ureasil-polyether films' surfaces, together with the recorded optical images. The comparison of the contact angle of the unloaded u-PEO500 and u-PPO400 films indicates that the value is significantly larger for u-PPO, which is consistent with its higher hydrophobic character. Due to its higher hydrophobic nature, u-PPO has lower affinity with water and therefore exhibits a higher contact-angle value. The incorporation of DMA induces changes in the contact angle values in both ureasil-polyether films.

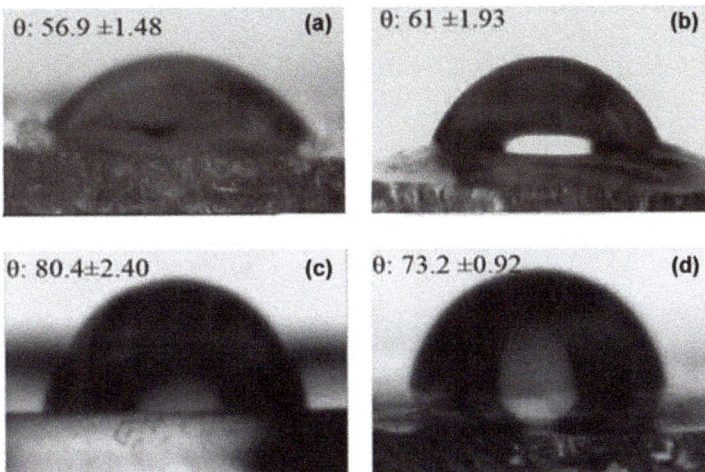

Figure 10. Contact-angle measurements on (**a**) unloaded u-PEO500; (**b**) DMA-loaded u-PEO500; (**c**) unloaded u-PPO400; (**d**) DMA-loaded u-PPO400. The results are expressed as the mean ± S.D. for n = 3 (replicates).

In the DMA-loaded u-PEO500 films, the contact angle increases as a result of the addition of DMA. According to the AFM results (Figure 4), no traces of DMA are found on the loaded u-PEO films surfaces. Nevertheless, as indicated by FT-IR (Figure 9), DMA does cause modifications in the u-PEO matrix by hindering the interactions between amide and ether oxygen from different polymer chains. If DMA and the poly(oxyethylene) moieties form complexes involving the ether oxygen, the presence of DMA may facilitate the elimination of water produced during the sol-gel reactions outside the hybrid film and reduce the number of reactive sites at surface locations in such a way that the resulting film surface exhibits reduced hydrophilicity.

In the case of u-PPO400 films, the addition of DMA to the polymer matrix diminishes the contact-angle value (i.e., increases the surface hydrophilicity). DMA is considered a hydrophobic molecule, with a low solubility in water. Nevertheless, the result is understandable considering that the structure of the u-PPO film is severely disrupted in the presence of DMA, with the formation of new crystalline phases. The increased hydrophilicity in DMA-loaded u-PPO400 film is bound to result in an increased bioadhesion of the film, with an increased probability of H-bonds formation on the film surface [13].

To evaluate surface modifications as a result of DMA release from u-PPO films, we immersed the samples in a solution containing 500 mL of medium (phosphate buffer 7.2 pH with 0.5% of procetyl AWS® (CRODA, Rawcliffe Bridge, UK)) in a manner similar to the drug release experiments reported in Figure 2, for 24 h, to ensure that little or no drug remained to be released. The images in Figure 11 were recorded over a (40,000 × 40,000) nm surface area of the u-PPO film after release of DMA into the medium. Figure 11a is the surface topography, and Figure 11b,c lateral force microscopy (LFM) images scanning from right to left (Figure 11b, recorded simultaneously with Figure 11a) and from left to right (Figure 11c, each line recorded as the tip travelled back along the line when recording Figure 11b). Figure 11d shows a height-contour profile along the continuous white line in Figure 11a. Figure 11e is a 3D representation of Figure 11a. The absence of frictional contrast in the LFM images Figure 11b,c confirms the chemical homogeneity of the film surface, i.e., the absence of DMA-related clusters on the film surface.

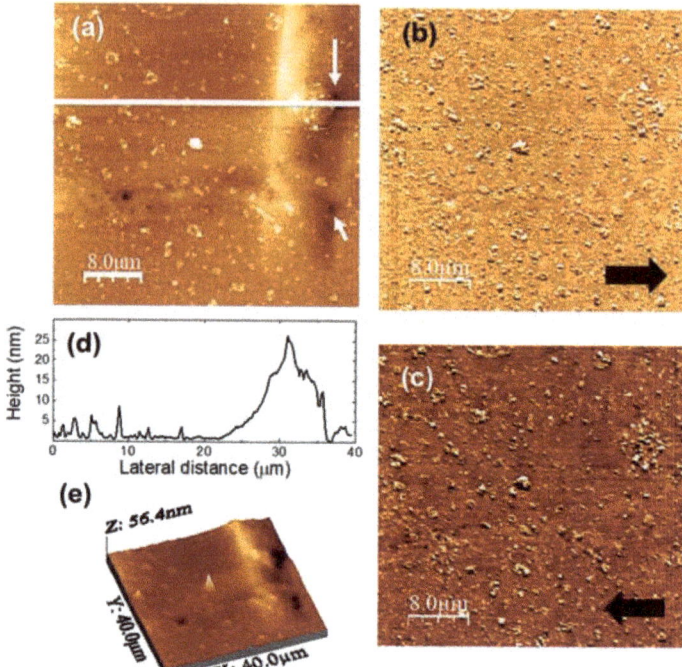

Figure 11. u-PPO500 film after DMA release in medium. (**a**) Contact-mode AFM topography; color-scale range: 35 nm. (**b,c**) Lateral force microscopy (LFM) images recorded (**b**) Scanning from left to right (simultaneously with (**a**)) and (**c**) from right to left. (**d**) Height-contours profile along the continuous white line in (**a**). (**e**) 3D representation of (**a**).

The surface morphology supports the information provided by the analysis of the release kinetics (Figure 2), according to which DMA release from u-PPO matrices mostly occurs by Fickian diffusion. No sign of erosion or swelling of the matrix substrate is apparent. The morphology is characterized by the presence of small clusters, typically lower than ~10 nm in high. The ~20 nm rectangular protrusion on the right-hand side of the image clearly indicates the previous location of a DMA crystal. Precisely at the protrusion sides, some small pores are apparent, such as those indicated by the white arrows in Figure 11a, also clearly noticeable in Figure 11e, which presumably form as the drug is released. Apparently, the molecular chains of the polymer matrix accompany the drug and re-form, as the drug diffuses outwards, leaving the matrix substrate.

4. Conclusions

The study we have carried out has allowed us to deepen our understanding of the nanostructural arrangements within ureasil-polyether matrices and how they interact with DMA drug molecules.

Key points disclosed by this work are the following:

- In u-PEO500, DMA occupies sites near the ether oxygen and remains absent from the film surface. The release kinetics of DMA from the u-PEO500 matrix fits well with Peppas' model; the formation of pores on u-PEO500 may enable the penetration of water inside the matrix and its swelling to facilitate the release process.
- In u-PPO400, new crystalline phases are formed upon loading with DMA, which show up as rounded clusters of ~30–100 nm in diameter, aligned along a well-defined direction. The study of the u-PPO400 matrix shows that it is structured in polymer

ropes, composed of strands, consisting of aligned clusters of ~180 nm in diameter, formed by aggregates of ureasil-nodes supercrystals. Hence, the new phases may grow between the strands, where ureasil nodes are absent. The formation of larger needle-shaped crystals has also been observed. The release kinetic of DMA from the u-PPO400 matrix fits well with Weibul model; when DMA release comes to an end, the matrix surface shows no trace of DMA and no evidence of erosion.

The results indicated above extend our current knowledge on ureasil-polyether, and DMA interactions within these matrices. In our opinion, the most relevant contribution of this work is the observation by means of UFM of aligned clusters of ureasil-node aggregates defining hybrid-polymer strands. UFM is a relatively new technique, and, in many areas, its full potential has not yet been explored. To our knowledge, no such structuring of uresil nodes has been observed before, and it should play a decisive role in many of the diverse applications of these hybrid polymer matrices. We believe it is likely that u-PEO500 is also structured into strands formed by aligned clusters of ureasil-node aggregates, but the presence of pores favors water penetration into these matrices, and swelling hinders their observation by AFM/UFM procedures under ambient conditions.

Author Contributions: Conceptualization, methodology, validation, and resources: J.A.O.-J., L.A.C. and M.T.C.; formal analysis: J.A.O.-J., A.L. and M.T.C.; investigation: J.A.O.-J., A.L., E.M.B. and M.T.C.; data curation: J.A.O.-J., A.L. and M.T.C.; writing—original draft preparation: M.T.C.; writing—review and editing, J.A.O.-J., L.A.C. and M.T.C.; visualization: J.A.O.-J. and M.T.C.; supervision, project administration and funding acquisition: L.A.C. and M.T.C. All authors have read and agreed to the published version of the manuscript.

Funding: The Coordenacão de Aperfeicoamento de Pessoal de Nível Superior-Brasil (CAPES), and the ERASMUS MUNDUS Programme, the ALBA Synchrotron Light Facility, the Programa Operativo FEDER de Castilla-La Mancha 2014-2020, and the Junta de Comunidades de Castilla-La Mancha (JCCM) are gratefully acknowledged for financial support.

Data Availability Statement: The data presented in this study are available on request from the corresponding author.

Acknowledgments: Marc Malfois is gratefully acknowledged for scientific advice regarding the authors' application for beamtime in ALBA, and for assistance during the SAXS measurements. Rodrigo Bastante is gratefully acknowledged for technical aid during the beamline time. Carlos Rivera Cavanillas (UCLM IRICA Instrumentation Service) is gratefully acknowledged for technical aid in XRD and FT-IR data acquisition.

Conflicts of Interest: The authors declare no conflict of interest. The funders had no role in the design of the study; in the collection, analyses, or interpretation of data; in the writing of the manuscript, or in the decision to publish the results.

References

1. Santilli, C.V.; Chiavacci, L.A.; Lopes, L.; Pulcinelli, S.H.; Oliveira, A.G. Controlled Drug Release from Ureasil−Polyether Hybrid Materials. *Chem. Mater.* **2009**, *21*, 463–467. [CrossRef]
2. Oshiro Junior, J.A.; Paiva Abucafy, M.; Berbel Manaia, E.; Lallo da Silva, B.; Chiari-Andreo, B.G.; Aparecida Chiavacci, L. Drug Delivery Systems Obtained from Silica Based Organic-Inorganic Hybrids. *Polymers* **2016**, *8*, 91. [CrossRef]
3. Souza, L.K.; Bruno, C.H.; Lopes, L.; Pulcinelli, S.H.; Santilli, C.V.; Chiavacci, L.A. Ureasil–polyether hybrid film-forming materials. *Colloids Surf. B Biointerfaces* **2013**, *101*, 156–161. [CrossRef] [PubMed]
4. De Zea Bermudez, V.; Carlos, L.D.; Alcácer, L. Sol-gel derived urea cross-linked organically modified silicates. 1. Room temperature mid-infrared spectra. *Chem. Mater.* **1999**, *11*, 569–580. [CrossRef]
5. Molina, E.F.; Jesus, C.R.N.; Chiavacci, L.A.; Pulcinelli, S.H.; Briois, V.; Santilli, C.V. Ureasil–polyether hybrid blend with tuneable hydrophilic/hydrophobic features based on U-PEO1900 and U-PPO400 mixtures. *J. Sol-Gel Sci. Technol.* **2014**, *70*, 317–328. [CrossRef]
6. Rafaella, M.B.; Maísa, S.d.O.; Kammila, M.N.C.; Mariana, R.S.; Karen, L.M.S.; Bolívar, P.G.d.L.D.; Teresa, C.; João, A.O.-J. Physicochemical Characterization of Bioactive Compounds in Nanocarriers. *Curr. Pharm. Des.* **2020**, *26*, 4163–4173. [CrossRef]
7. Cuberes, M.T. Mechanical Diode-Based Ultrasonic Atomic Force Microscopies. In *Applied Scanning Probe Methods XI. Nanoscience and Technology*; Springer: Berlin, Heidelberg, 2009. [CrossRef]

8. Ma, C.; Arnold, W. Nanoscale ultrasonic subsurface imaging with atomic force microscopy. *J. Appl. Phys.* **2020**, *128*, 180901. [CrossRef]
9. Sitterberg, J.; Özcetin, A.; Ehrhardt, C.; Bakowsky, U. Utilising atomic force microscopy for the characterisation of nanoscale drug delivery systems. *Eur. J. Pharm. Biopharm.* **2010**, *74*, 2–13. [CrossRef]
10. Lopes, L.; Molina, E.F.; Chiavacci, L.A.; Santilli, C.V.; Briois, V.; Pulcinelli, S.H. Drug-matrix interaction of sodium diclofenac incorporated into ureasil-poly(ethylene oxide) hybrid materials. *RSC Adv.* **2012**, *2*, 5629–5636. [CrossRef]
11. Molina, E.F.; Pulcinelli, S.H.; Santilli, C.V.; Blanchandin, S.; Briois, V. Controlled Cisplatin Delivery from Ureasil−PEO1900 Hybrid Matrix. *J. Phys. Chem. B* **2010**, *114*, 3461–3466. [CrossRef]
12. Molina, E.F.; Santilli, C.V.; Pulcinelli, S.H.; Blanchandin, S.; Baudelet, F.; Briois, V. Multi-spectroscopic monitoring of cisplatin-derived species delivery from ureasil polyether hybrid matrix. *Phase Transit.* **2011**, *84*, 687–699. [CrossRef]
13. Oshiro Junior, J.A.; Carvalho, F.C.; Soares, C.P.; Chorilli, M.; Chiavacci, L.A. Development of Cutaneous Bioadhesive Ureasil-Polyether Hybrid Films. *Int. J. Polym. Sci.* **2015**, *2015*, 1–7. [CrossRef]
14. Mariano, G.H.; Gomes de Sá, L.G.; Carmo da Silva, E.M.; Santos, M.A.; Cardozo Fh, J.L.; Lira, B.O.V.; Barbosa, E.A.; Araujo, A.R.; Leite, J.R.S.A.; Ramada, M.H.S.; et al. Characterization of novel human intragenic antimicrobial peptides, incorporation and release studies from ureasil-polyether hybrid matrix. *Mater. Sci. Eng. C* **2021**, *119*, 111581. [CrossRef]
15. Gómez-Gaete, C.; Tsapis, N.; Besnard, M.; Bochot, A.; Fattal, E. Encapsulation of dexamethasone into biodegradable polymeric nanoparticles. *Int. J. Pharm.* **2007**, *331*, 153–159. [CrossRef]
16. Zilberman, M.; Schwade, N.D.; Meidell, R.S.; Eberhart, R.C. Structured drug-loaded bioresorbable films for support structures. *J. Biomater. Sci. Polym. Ed.* **2001**, *12*, 875–892. [CrossRef]
17. Zilberman, M. Dexamethasone loaded bioresorbable films used in medical support devices: Structure, degradation, crystallinity and drug release. *Acta Biomater.* **2005**, *1*, 615–624. [CrossRef]
18. Rodrigues, G.; Silva, D.; Ayres, E.; Orefice, R.L.; Moura, S.A.L.; Cara, D.C.; Da, A.; Cunha, S. Controlled release of dexamethasone acetate from biodegradable and biocompatible polyurethane and polyurethane nanocomposite. *J. Drug Target.* **2009**, *17*, 374–383. [CrossRef]
19. Noreen, S.; Maqbool, I.; Madni, A. Dexamethasone: Therapeutic potential, risks, and future projection during COVID-19 pandemic. *Eur. J. Pharmacol.* **2021**, *894*, 173854. [CrossRef]
20. Bruschi, M.L. Mathematical models of drug release. In *Strategies to Modify the Drug Release from Pharmaceutical Systems*; Bruschi, M.L., Ed.; Woodhead Publishing Limited (Elsevier Ltd.): Cambridge, UK, 2015.
21. Wojdyr, M. Fityk: A gneral-purpose peak fitting program. *J. Appl. Crystallogr.* **2010**, *43*, 1126–1128. [CrossRef]
22. Toby, B.H.; Von Dreele, R.B. GSAS-II: The genesis of a modern open-source all purpose crystallography software package. *J. Appl. Crystallogr.* **2013**, *46*, 544–549. [CrossRef]
23. Horcas, I.; Fernández, R.; Gómez-Rodríguez, J.M.; Colchero, J.; Gómez-Herrero, J.; Baro, A.M. WSXM: A software for scanning probe microscopy and a tool for nanotechnology. *Rev. Sci. Instrum.* **2007**, *78*, 013705. [CrossRef]
24. Carbinatto, F.M.; de Castro, A.D.; Evangelista, R.C.; Cury, B.S.F. Insights into the swelling process and drug release mechanisms from cross-linked pectin/high amylose starch matrices. *Asian J. Pharm. Sci.* **2014**, *9*, 27–34. [CrossRef]
25. Carvalho, F.C.; Campos, M.L.; Peccinini, R.G.; Gremião, M.P.D. Nasal administration of liquid crystal precursor mucoadhesive vehicle as an alternative antiretroviral therapy. *Eur. J. Pharm. Biopharm.* **2013**, *84*, 219–227. [CrossRef]
26. Costa, P.; Sousa Lobo, J.M. Modeling and comparison of dissolution profiles. *Eur. J. Pharm. Sci.* **2001**, *13*, 123–133. [CrossRef]
27. Sun, W.; Zhang, N.; Li, X. Release mechanism studies on TFu nanoparticles-in-microparticles system. *Colloids Surf. B Biointerfaces* **2012**, *95*, 115–120. [CrossRef]
28. Siepmann, J.; Peppas, N.A. Higuchi equation: Derivation, applications, use and misuse. *Int. J. Pharm.* **2011**, *418*, 6–12. [CrossRef]
29. Korsmeyer, R.W.; Gurny, R.; Doelker, E.; Buri, P.; Peppas, N.A. Mechanisms of solute release from porous hydrophilic polymers. *Int. J. Pharm.* **1983**, *15*, 25–35. [CrossRef]
30. Papadopoulou, V.; Kosmidis, K.; Vlachou, M.; Macheras, P. On the use of the Weibull function for the discernment of drug release mechanisms. *Int. J. Pharm.* **2006**, *309*, 44–50. [CrossRef]
31. Oshiro, J.A.; Nasser, N.J.; Chiari-Andreó, B.G.; Cuberes, M.T.; Chiavacci, L.A. Study of triamcinolone release and mucoadhesive properties of macroporous hybrid films for oral disease treatment. *Biomed. Phys. Eng. Express* **2018**, *4*. [CrossRef]
32. Marino, S.; Joshi, G.M.; Lusuardi, A.; Cuberes, M.T. Ultrasonic force microscopy on poly(vinyl alcohol)/SrTiO3 nano-perovskites hybrid films. *Ultramicroscopy* **2014**, *142*. [CrossRef]
33. Kuhnert-Brandstätter, M.; Gasser, P. Solvates and polymorphic modifications of steroid hormones III. *Microchem. J.* **1971**, *16*, 590–601. [CrossRef]
34. Silva, R.P.d.; Ambrósio, M.F.S.; Piovesan, L.A.; Freitas, M.C.R.; Aguiar, D.L.M.d.; Horta, B.A.C.; Epprecht, E.K.; San Gil, R.A.d.S.; Visentin, L.d.C. New Polymorph Form of Dexamethasone Acetate. *J. Pharm. Sci.* **2018**, *107*, 672–681. [CrossRef] [PubMed]
35. Terzis, A.; Theophanides, T. The crystal and molecular structure of 9[alpha]-fluoro-16[alpha]-methyl-11[beta],17,21-trihydroxy-1,4-pregnadiene-3,20-dione 21-acetate monohydrate. *Acta Crystallogr. Sect. B* **1975**, *31*, 796–801. [CrossRef]
36. Mallet, F.; Petit, S.; Petit, M.N.; Cardinaël, P.; Billot, P.; Lafont, S.; Coquerel, G. Solvent exchange between dimethylsulfoxide and water in the dexamethasone acetate structure. *J. Phys.* **2001**, *11*, Pr10-253–Pr210-259. [CrossRef]
37. Delage, S.; Couvrat, N.; Sanselme, M.; Cartigny, Y.; Coquerel, G. Stability of solid phases in the dexamethasone acetate/water system Stability of solid phases in the dexamathasone acetate/water system. *MATEC Web Conf.* **2013**. [CrossRef]

38. Mallet, F.; Petit, S.; Lafont, S.; Billot, P.; Lemarchand, D.; Coquerel, G. Solvent exchanges among molecular compounds. *J. Therm. Anal. Calorim.* **2003**, *73*, 459–471. [CrossRef]
39. Artusio, F.; Pisano, R. Surface-induced crystallization of pharmaceuticals and biopharmaceuticals: A review. *Int. J. Pharm.* **2018**, *547*, 190–208. [CrossRef]
40. Oshiro Junior, J.A.; Shiota, L.M.; Chiavacci, L.A. Desenvolvimento de Formadores de Filmes Poliméricos Orgânico-Inorgânico para Liberação Controlada de Fármacos e Tratamento de Feridas. *Matéria* **2014**, *19*, 24–32.
41. Stanley, E.; Litt, M. Crystal structure of d,l-poly(propylene oxide). *J. Polym. Sci.* **1960**, *43*, 453–458. [CrossRef]
42. Cesari, B.M.; Perego, G. The Crystal Structure of Isotactic Poly (propylene oxide). *Die Makromol. Chem. Macromol. Chem. Phys.* **1966**, *94*, 194–204. [CrossRef]
43. Molina, C.; Dahmouche, K.; Hammer, P.; Bermudez, V.D.Z.; Carlos, L.D.; Ferrari, M.; Montagna, M.; Gonçalves, R.R.; Oliveira, L.; Edwards, H.; et al. Structure and Properties of Ti4+-Ureasil Organic-Inorganic Hybrids. *J. Braz. Chem. Soc.* **2006**, *17*, 443–452. [CrossRef]
44. Dahmouche, K.; Santilli, C.V.; Pulcinelli, S.H.; Craievich, A.F. Small-Angle X-ray Scattering Study of Sol−Gel-Derived Siloxane−PEG and Siloxane−PPG Hybrid Materials. *J. Phys. Chem. B* **1999**, *103*, 4937–4942. [CrossRef]
45. Cohen, E.M. Dexamethasone. *Anal. Profiles Drug Subst. Excip.* **1973**, *2*, 163–197. [CrossRef]
46. Fernandes, M.; Barbosa, P.C.; Silva, M.M.; Smith, M.J.; Zea, V.D. Di-ureasil hybrids doped with LiBF 4: Spectroscopic study of the ionic interactions and hydrogen bonding. *Mater. Chem. Phys.* **2011**, *129*, 385–393. [CrossRef]

MDPI
St. Alban-Anlage 66
4052 Basel
Switzerland
Tel. +41 61 683 77 34
Fax +41 61 302 89 18
www.mdpi.com

Nanomaterials Editorial Office
E-mail: nanomaterials@mdpi.com
www.mdpi.com/journal/nanomaterials

www.ingramcontent.com/pod-product-compliance
Lightning Source LLC
LaVergne TN
LVHW070620100526
838202LV00012B/692